Bathrooms

Bathrooms

A Professional's Illustrated Design and Remodeling Guide

Chase Powers

McGraw-Hill

New York San Francisco Washington, D.C. Auckland Bogotá
Caracas Lisbon London Madrid Mexico City Milan
Montreal New Delhi San Juan Singapore
Sydney Tokyo Toronto

Library of Congress Cataloging-in-Publication Data

Powers, Chase M.
    Bathrooms : a professional's illustrated design and remodeling
guide / Chase M. Powers.
        p.   cm.
      Includes index.

CIP

McGraw-Hill

A Division of The **McGraw·Hill** *Companies*

1 2 3 4 5 6 7 8 9 0 DOC/DOC 9 0 2 1 0 9 8 7

ISBN 0-07-008629-x (HC)
ISBN 0-07-008628-1 (PBK)

*The sponsoring editor for this book was Zoe G. Foundotos, the editing supervisor
was Sally Glover, and the production supervisor was Sherri Souffrance. It was set
in the AN1 design in Souvenir by Michele Bettermann and Michele Pridmore of
McGraw-Hill's desktop publishing department in Hightstown, N.J.*

Printed and bound by R. R. Donnelley & Sons Company.

 This book is printed on recycled, acid-free paper containing a minimum of
50% recycled, de-inked fiber.

This book is dedicated to Victoria Roberts and others who know who they are.

Contents

Acknowledgments

I'd like to thank Victoria Roberts for her efforts in helping me obtain art for this book, and to thank the following companies for being so helpful in providing art and illustrations that make this book what it is. The names appear in the order of the materials used:

Pella

WeatherShield

Andersen Windows, Inc.

American Chinaware

St. Thomas Creations

Ackermann

Wilsonart

Dal-Tile

Hurd

Vellux

Eljer

Universal Rundel

Fran Pagurko

Images International

Acknowledgments

Lone Wolf Enterprises, Ltd.

Sanderson Garden Collection

Americh

Waterworks

Porcher Ltd., a Division of American Standard, Inc.

Hansa America

Monet Faucets

JADO

Moen

Nutone

Armstrong World Industries, Inc.

Georgia Pacific

Style-Mark

Pergo

Simpson Door Company

CraftMaster Marketing

Caradco Wood Windows & Patio Doors

Liz King

Paris Ceramics

Vista Window Film

Wellborn Cabinets

Merillat Industries

Plain & Fancy Cabinets

Horton Brass

Introduction

Bathroom remodeling is a very lucrative business. In fact, bathroom and kitchen remodeling are the two most popular forms of remodeling, and they return the biggest financial yields. If you are a bathroom remodeler, you are in a good business. After more than two decades in the remodeling business, I've seen many changes, and contractors must change too or be left behind by their competitors. The fact that you are reading this book indicates that you are wise. You must further your knowledge to prosper in the growing field of remodeling, and reading is an excellent way to keep up-to-date on new products and techniques. This book is full of ideas, suggestions, and directions to keep you a step or two in front of average remodelers.

Whether you are just starting in the business of bathroom remodeling or are a seasoned veteran, you are sure to find information here that you can put to use on your next job. We are going to take bathroom remodeling from start to finish, and I do mean from start to finish. The discussion will open with the planning and design phase. We will look at the elements that combine to make beautiful bathrooms that you will be happy to have your name associated with. Then we will talk about cost estimates. As you may know, the accuracy of your estimates will make or break you in the business world, and Chapter 3 will give you the hard facts that you need to be competitive and profitable.

Most contractors rely on subcontractors for at least some part of remodeling projects, and many contractors work only with subcontractors. This subject will be dealt with, in detail, in Chapter 4. Before we get into hands-on work, we will also cover saving money on labor and materials and how making quick decisions can turn a dream job into a nightmare.

Rip-out work is a routine part of most bathroom remodeling, but there are right ways and wrong ways of doing it. See Chapter 7 for complete instructions on how to make the most of a rip-out. Plumbing is obviously a major part of bathroom remodeling, and there are two chapters here that are dedicated to this subject. Electrical and HVAC aspects of the job are also covered. The mechanical trades are a big portion of a bathroom job, and you will find plenty of good reading on the subject.

Floors, walls, ceilings, cabinets, and counters are also significant aspects of a remodeling job. As you might expect, all of these topics are talked about from both a practical and artistic point of view. You will see how installations are done and what the results of good workmanship should look like. There is even a chapter on accessories to enhance the appearance of a bathroom.

Is there anything of importance that is not covered in this book? I don't think so, and I believe you will agree with me. Could you find a better book on the subject? Perhaps, but I don't know where, and I think that you will agree that this book is a professional tool rather than just casual reading material. I've written this book for you, the professional. You will not have to wade through a lot of step-by-step instructions like you might find in some books that are intended for a different audience. This book is written by an experienced professional for professionals. You will gain a new look on bathroom remodeling before you are done perusing these pages. There are also hundreds of illustrations to give you ideas. Fresh thoughts are often the best means of winning more bids and making happy customers, and this thought-provoking source should do much to propel your career in bathroom remodeling.

Take a few moments to look over the content here. You will find the writing reader-friendly and the illustrations outstanding. I hope you enjoy reading it as much as I enjoyed writing it, and I'm sure you will leave the book with a feeling of fulfillment.

Chapter summaries

Chapter 1: The right plan

This chapter will help contractors settle on a perfect plan for bathroom design and remodeling. Contractors can use this chapter to sit down with customers and arrive at a working plan that enjoys the benefit of over 23 years of my hands-on experience.

Chapter 2: Beautiful bathroom designs

Chapter 1 shows contractors and their customers how to develop a solid plan for designing and remodeling bathrooms, and Chapter 2 picks up where the previous chapter leaves off. A large number of photographs help readers visualize a host of decorative bathroom designs and features. Basically, this chapter is a catalog of winning bathroom designs.

Chapter 3: Cost estimates

Estimating the cost of bathroom remodeling can be very tricky. With over two decades of experience under my belt, I've learned how to expect the unexpected. This chapter will prepare contractors for making accurate bids that they will not regret. I will point out risks, likely problems, and general estimating techniques in this chapter.

Chapter 4: Subcontractors' bids

Chapter 4 shows contractors how to maximize their use of subcontractors. I teach readers how to solicit bids, evaluate subcontractors, drive a hard bargain, and deal with subcontractors from start to finish. Subcontractors are the heart of any general-contracting business, so this chapter is must-read material.

Chapter 5: Savings negotiations

Everyone likes to save money. For general contractors, saving money can be as easy as asking for extra discounts. Over the years I've learned a myriad of ways to cut my supply costs to the bone, and I share my money-saving secrets in this chapter. It also highlight ways to get great deals on everything from wallpaper to bathtubs.

Chapter 6: Quick decisions

A remodeler's biggest mistake can be making quick decisions. Builders who have to make on-the-spot decisions can ruin their relationships with customers. Building and remodeling is filled with surprises that can put contractors into a world of trouble. I learned a long time ago to think long and hard before I speak to a customer. This chapter educates readers on the risks of talking off the tops of their heads. I show contractors how to take control of situations, evaluate circumstances, and communicate with customers in an effective, efficient manner.

Chapter 7: Ripping out an existing bathroom

Before a new bathroom can replace an existing one, the old bathroom has to be ripped out. Removing fixtures, flooring, and other elements of a bathroom can be simple, but it can also harbor a host of dangers. There is an art to successful rip-outs, and this chapter shows contractors how to orchestrate a desirable demolition. I cover all aspects of a remarkable rip-out. Topics include everything from dust control and debris removal to alternative options for homeowners who must get by with just one bathroom while it's being remodeled.

Chapter 8: Selecting the best plumbing fixtures

When it comes to helping customers choose the right plumbing fixtures, the process can be frustrating, confusing, and awkward. Many general contractors have a limited knowledge of what is available for their customers when it comes to lavatories, toilets, bidets, showers, and bathtubs. As a master plumber, I show contractors—through both words and photographs—how to understand all of the options available to them and their customers. I cover all major types of plumbing fixtures in a way that is both easy to understand and comparative in nature.

Chapter 9: Plumbing points to ponder

Existing pipes and fittings can give a plumber a fit when remodeling a bathroom. As bad as the basics of plumbing can be for a plumber, they can be much worse for a remodeler. In a perfect world, every remodeler would have a plumber evaluate each job before a bid is made. In reality, this is not always the case. I use this chapter to show contractors what to look for in terms of trouble spots. Telling contractors what to expect when their plumbers show up will produce numerous advantages. Contractors will be able to prepare their customers for what's to come. When a contractor is knowledgeable about plumbing procedures, it is easier to coordinate and supervise work to be complete. This chapter gives readers the benefit of my many years of experience as a builder, remodeler, and master plumber.

Chapter 10: Electrical considerations

Electrical considerations in bathroom design and remodeling can run the gamut from the removal of electric heat to the installation of make-up lights to illuminating a new dressing area. Most electrical work in a bathroom job is fairly simple, but there can be problems. If a contractor knows what to look for when giving an estimate, a lot of trouble can be avoided. From a design point of view, planning efficient electrical provisions pays big dividends. This chapter covers all major aspects of electrical codes and considerations for bathrooms. Light fixtures for bathrooms are also be covered in this chapter.

Chapter 11: HVAC aspects of a job

HVAC issues can create complications in a bathroom. By nature, most bathrooms are small. This can make it difficult to place heating, ventilation, and air-conditioning facilities. I use this chapter to show readers how to design optimum use of HVAC systems. In addition, I detail the options for changes, additions, and deletions of HVAC systems in a bathroom.

Chapter 12: Flooring options

Flooring options for bathrooms are limited, at least in theory and by code requirements, to nonabsorbent materials. This usually means a tile or vinyl flooring. Some bathrooms are outfitted with carpeting, and I've seen some with hardwood flooring. There are times when carpeting is a good idea, especially when the user of the room is physically challenged and requires the use of crutches. Maneuvering around a wet bathroom floor made of tile when you must rely on crutches for mobilization can be quite dangerous. This chapter investigates ceramic tile, quarry tile, sheet vinyl, vinyl tiles, carpeting, and other flooring possibilities.

Chapter 13: Walls and ceilings

Walls and ceilings in bathrooms can cause countless headaches, whether a contractor is planning the construction of a new bathroom, bidding a remodeling job, or just dealing with existing conditions. Moisture is always an issue in bathrooms. Improper ventilation can lead to mildew, rot, and other situations that will ruin the best of bathrooms. Customers often have trouble choosing a suitable wallcovering for their bathrooms. Some options include ceramic tile, wallpaper, and paint. Decorative stenciling is also a possibility. Installing moisture-resistant drywall should be a given in a bathroom, but some builders don't bother with this type of protection. This chapter provides a comprehensive guide to walls and ceilings in bathrooms.

Chapter 14: Cabinets and counters

Not all bathrooms utilize cabinets and counters. Pedestal and wall-hung lavatories can be used in place of vanities. However, the use of a vanity provides a homeowner with extended storage space and more counter space. Should a contractor recommend a fashionable pedestal lavatory or a functional vanity? This chapter answers this question and many more. We review the types of cabinets and counters that are available, as well as the hidden problems that can be associated with certain types of cabinets. For example, if a vanity with drawers is going to be used, the plumber must be aware of where to rough in pipes to avoid contact with the drawers. Many line drawings accent this chapter to show the specifics of various types of cabinets. I also talk about the different options for lavatories being used with vanities, such as cultured-marble tops, drop-in lavatories, self-rimming lavatories, and other options.

Chapter 15: Accessories to enhance a bathroom

This chapter is a smorgasbord of ideas for creative, attractive, and functional bathroom accessories. Photographs and line drawings make this a very visual chapter.

Glossary

Here we identify and define a large number of terms that are used in bathroom design, construction, and remodeling.

The right plan

CHOOSING the right plan for a remodeling job is essential. Low prices, quality work, and many other elements contribute to a contractor's ability to get work, but helping customers plan their work is almost always the first step in winning a job. Some customers know exactly what they want when they call contractors for bids, but there are many more who need help in designing their new bathroom. It's not just the new bathroom that has to be decided upon. Much of the planning has to do with demolition of an existing bathroom when working with remodeling. Whether you are laying out a bathroom for new construction or a remodeling project, the planning that goes into the project before bidding is crucial to the success of the job. If you, as a contractor, can help your customers with this critical phase of their jobs, you are much more likely to win the job, even if your prices are a little higher than your competitor's.

Homeowners and people having new homes built often buy blueprints from stock-blueprint books. Architects are sometimes hired, but their fees can be too high for many customers. Another option for someone interested in bathroom design is a plumbing contractor. I've been in the construction and remodeling business for about 23 years. My work has included being a general contractor, a remodeling contractor, and a plumbing contractor. At one time, I was building upwards of 60 single-family homes a year. My remodeling volume has been extensive, especially in the field of bathroom and kitchen remodeling. After more than two decades of working with customers and bathrooms, I've learned a lot.

A few of my customers have come to me with plans drawn by local architects. Many others have drawn their own sketches of what they would like their jobs to look like. This is a topic we will discuss more deeply in a few minutes. A good number of people have approached me with line drawings that they have seen in stock blueprint magazines. But far and away, most of my customers have come to me with random ideas and a lot of questions. Customers who don't have a definitive plan are what you are most likely to encounter in your business.

Your skill and creativity is needed to help customers who either don't know what they want or who have not been able to put their dreams down on paper. Even when customers do know what they want, it may be up to you to make suggestions to correct items that you recognize as potential mistakes. Don't be afraid to interject your opinions. Most customers welcome the opinions of professionals. Of course, you should be tactful when telling potential customers that their ideas stink. Let your thoughts be known in a professional manner that the customer will see as help, rather than ridicule. You will certainly run into some customers who have ideas that you would not want your good business name associated with. There are ways to work around such problems. Professional advice is one of the most effective.

Rough sketches

Rough sketches are good tools when designing a bathroom. However, for a sketch to be of much use, it should be drawn to scale. Few customers pay much attention to scale drawings when they are putting their thoughts on paper. This is okay, so long as the customer is made aware that a drawing that is not made to scale may not work out well. Let your customers doodle all they want. Encourage customers to draw a variety of layouts for you to work with. The more information you can obtain from the customer, the better off you will be. Once you have their renderings, you can put the drawings into perspective by redrawing them to scale. You can do this on a computer or on graph paper. How you arrive at the finished sketch is up to you, but it must be to scale if it is to be used effectively.

If you've been in the business for long, you must have had people give you drawings that showed all sorts of stuff arranged in a new bathroom. Closets, whirlpools, large vanities, and a host of other elements will fit in a rectangle on paper just fine. However, putting the same items in a real bathroom may not work out at all. A lot of bathrooms are small. Dimensions of five feet by eight feet are common. You can only get so much into such a space. This is why the drawings have to be scaled out. It doesn't matter what scale you use, so long as it is consistent throughout the plan.

Stock blueprints

Most bookstores sell books and magazines that have drawings of floor plans in them. Stock blueprints can be ordered from the floor plans, and this is an economical way to obtain detailed blueprints. However, many of the plans wind up being customized before construction or remodeling begins. I maintain an extensive library of books with floor plans in them. When customers come to me in search of ideas, I give them a variety of books to look through. It's common for customers who sit down with me to take several bathroom layouts and compile special elements of each to create a single plan. This type of work is extremely effective for me and the customers.

I've used stock blueprints on many occasions. In fact, most of my personal homes have started out as stock blueprints. Altering the prints can alter structural conditions, so you must be sure not to make changes that will affect the structural integrity of your job. If necessary, hire an architect to review proposed changes to make certain that they are safe. Cosmetic changes can be made without the need of architectural approval. Your job will probably become much easier if you invest in a wide variety of plan books to show your customers.

Experience

Experience is one of the best assets you can have when it comes to the designing of a bathroom. Unfortunately, most contractors have to make a lot of mistakes to gain the experience needed to create dramatic bathroom designs (Fig. 1-1). Yes, I've made the mistakes, but hopefully, I can shorten your learning curve with this book and prevent you from falling into the traps that I have over the years. There is no substitute for experience. Few books tell you about finding bees nesting in walls that you are in the process of ripping out. Only some books or magazines tell you how to deal with specialty fixtures, such as gold faucets. I once had a plumber who destroyed two gold lavatory faucets at a cost of about $5,000 to me. The damage was done because of the soft metal and the plumber's inexperience. I had

Figure 1-1

The windows in this bathroom make it quite a showplace. Courtesy of Weathershield

to foot the bill and appease the customer. It's not easy getting started in business, and staying in it successfully is even more difficult. Until you gain the experience you are at a disadvantage, and you can't get the experience until you have survived. It is the proverbial double-edged sword. We're talking about this experience stuff for a reason. I want you to pay attention to what we are about to get into. Don't skim the pages; read them. I've been there and paid the price of ignorance; you don't have to. If you will absorb the information among these pages, you should be able to do your job better, more profitably, and for a much longer duration.

Square rooms

Square rooms don't usually make for an efficient bathroom. There is generally a lot of wasted space in a square room. Rectangles (Fig. 1-2),

L-shapes, and other layouts (Fig. 1-3) offer much more potential for the highest and best use of square footage. If you design a bathroom as a square, there is going to be a lot of open space in the middle of the room. The space might be okay, but it could be put to better use. With some exceptions, square rooms should be avoided.

What happens if you are working with an existing home where the bathroom is in a large, square room? Is there anything that you can do to make the room more efficient? There may be, depending on the room dimensions and the desires of your customer. It's possible that you can expand the bathroom (Fig. 1-4). This is often possible if other

Figure 1-2

This bathroom makes good use of space. Courtesy of Pella

Figure 1-3

Windows and an enormous space are what make this bathroom outstanding. Courtesy of Andersen Windows, Inc.

Figure 1-4

Expanding the size of a bathroom is often possible, and it's practical when other remodeling is already taking place. Courtesy of Andersen Windows, Inc.

remodeling is going on at the same time. Let me relate a story from my past to you about an old farmhouse that had a large, square bathroom to be remodeled.

It's been many years, but I remember pretty clearly my trip to a farmhouse for a bathroom estimate. The old house had been purchased recently by a new owner. The bathroom was ancient and very large. Due to the shape of the room, it seemed like there was going to be substantial wasted space. There was an old, clawfoot tub that would be refinished and used in the new bathroom. The old lavatory and toilet would be discarded and new ones would be installed. I planned to relocate existing piping to allow for a more spacious fixture placement. The homeowners and I discussed a large, double-bowl lavatory on a vanity. The cultured marble top would extend past the vanity to serve as a makeup area. This would help to fill up one wall of the room. We talked about installing a bidet, but a decision was made not to. The end product would be a somewhat typical bathroom in a space large enough to be a bedroom. Then I came up with an idea for the middle of the room. I suggested that the customers have my crews install a large whirlpool tub in an island enclosure. After an evening of thought, the homeowners called me with great excitement. They loved the idea. My draftsman put together some scaled drawings for the room and we added some other nice features, such as closet space, magazine racks, specialty lighting, and so forth. By the time we finished the design, the bathroom looked perfectly proportioned.

Even with the whirlpool in the middle of the room, there was still going to be a lot of space around it. We solved this problem by building a tile enclosure that came down and out like steps. The job looked terrific on paper. Fortunately, the finished job was not a disappointment. The homeowners were ecstatic and referred me to some of their friends. A job that started out looking like a loser turned out to be a big winner.

It often takes creative thought to get around problems in remodeling and sometimes even in new construction. If you look and think hard enough, there are usually acceptable answers waiting to be found. Don't give up, and don't accept matters at face value. Keep searching for suitable solutions until you find at least one, and hopefully more, that will make the job you are preparing to do a stand-out job. Your reputation will grow and shine as you demonstrate an ability to go further than average contractors with your design and implementation skills.

Rectangular rooms

Rectangular rooms are where most bathrooms are housed. These rooms lend themselves to cost-effective constructions and a reasonably good use of space. Unfortunately, a majority of the rooms are too small to allow a lot of creative control when remodeling. If you are

Figure 1-5

Window designs and types can add space to a bathroom or make a room appear more spacious.
Courtesy of Andersen Windows, Inc.

Figure 1-6

A pedestal lavatory. Courtesy of American Chinaware

faced with a standard bathroom, say one with dimensions of five by eight, you've got your work cut out for you. It's easy enough to swap out the fixtures, but giving the room some elegant appeal will be challenging. You might be able to do this, however, with a cantilever, a window grouping, or some other type of adjustment (Fig. 1-5). If you cannot make a physical change in the size of the room, your options are limited.

Can you enlarge the existing room? It's usually difficult to accomplish such a goal without major expense. Since many bathrooms are located on outside walls, you may be able to expand outward with an addition. Maybe you can steal some space from a walk-in closet in an adjoining room. Is the homeowner willing to shrink a bedroom to enlarge the bathroom? It's a possibility on many jobs. Can you obtain enough extra room to make the work worthwhile? Whenever you are talking about an expansion project, there will be many facets to consider.

Bigger is not always better in bathrooms. If the room is too large, the space will look empty. Matching fixtures and fixture styles to the space that is available is one of the best ways to arrive at an appealing job. This can be done with corner toilets, pedestal lavatories (Fig. 1-6), showers, corner showers, garden tubs, and a number of other fixture options (Fig. 1-7). In order to reach a desirable design, you must have ideas. One of the best ways to maintain fresh ideas is to read a lot. Books like this one and many magazines can be of great help in keeping your creative juices flowing. Remodelers who don't invest the time and energy to learn new techniques and styles often get left behind in the race for profits.

Figure 1-7

Taking advantage of various tub sizes and shapes allows more creativity in bathroom design. Courtesy of Andersen Windows, Inc.

L-shaped rooms

L-shaped rooms were often used to accommodate bathrooms in older homes. They have not been as popular in recent years in the areas where I've worked. One problem with the L-shape design is that the toilet often ends up being in the alcove of the L-shape. This presents an almost tunnel-like effect that can feel quite confining. I've remodeled a number of bathrooms where the toilets were stuck in a little square at the end of the bathtub. Technically, this is a good use of space, but aesthetically it is not a good move. There are, however, ways to overcome such existing conditions without extreme measures.

Assume that you have a bathroom where the toilet is set in a three-sided recess. What will you do to make the job more attractive? You might reconfigure the room. If changing the room design is completely out of the question, you must turn to other methods. Assuming that one of the walls enclosing the toilet is an outside wall, you could have a nice window to lighten up the alcove. The bathtub could be removed and replaced with a shower stall. This would allow the interior partition on one side of the toilet to be removed, giving the room a more spacious look and feel. If the tub is going to stay but will not have shower walls around it, you could cut out a section of the wall that is common between the tub and the toilet. Create a half-wall and allow the open concept to brighten up the room. Using creative ideas can result in some good-looking bathrooms without a lot of major remodeling.

Another problem with some L-shaped bathrooms is the location of the lavatory. It is frequently about midway down the wall opposite of the tub. Since so many of these bathrooms are tiny, the space to walk between the lavatory and the tub, to get to the toilet, can be very tight. A lot of the lavatories are old wall-hung units. Due to space constraints, putting a vanity and a top in such a bathroom can be all but impossible. One solution is a corner vanity and lavatory. By relocating the lavatory, you open up the walkway in the bathroom considerably.

Custom-made cabinets can be used to enhance a small bathroom where stock sizes are just too big. Replacing a tub or tub-shower combination with a standard shower stall can also improve the room size. Eliminating an existing linen closet might buy you some space. You and your customers will have to give a project plenty of thought if your goal is to end up with an outstanding room.

Odd-shaped rooms

Odd-shaped rooms come along every now and then. You can find them in new homes and in old houses. Rooms that don't have routine shapes can make life interesting for a remodeling contractor. With the right touch, a room that seems odd at first can be made to look spectacular (Fig. 1-8). Fortunately, there are a number of plumbing

Figure 1-8

The windows and design features of this bathroom make it remarkable.
Courtesy of Andersen Windows, Inc.

Figure 1-9

The corner lavatory here is a space-saver. <small>Courtesy St. Thomas Creations</small>

fixtures available for angles, curves, and corners (Fig. 1-9). And it's possible to have some plumbing fixtures, such as shower bases, made right on the site. You can do a lot with a tile shower in any shape. Money is usually an issue that prohibits total freedom of creativity. But when money is not a concern, the options are almost unlimited.

Design basics

Design basics for bathrooms are not difficult to grasp. Simple bathrooms have only three fixtures in them. Other fixtures are sometimes present, but the general concept of where and how to install the fixtures is not complex. There are, however, some rules-of-thumb to follow for a pleasant design. There are also code issues to keep in mind when deciding on fixture placement.

Before you can draw plans for a new bathroom, you have to know what your customers want. This can be difficult, since customers often don't know themselves what they want. You may very well have to educate your customers before you can bid their jobs. This is not as hard to do as you might think. Send your customers to some fixture showrooms. Let the people look through books of bathroom designs and select elements from various ones that they like. Be prepared for the customers to require a substantial amount of time and changes before they arrive at a final goal.

There's not much point in putting a lot of hours into an estimate until you have your customers settled into a plan that they love. Expect customers to change their minds frequently during the planning stages and even during the construction or remodeling stages. Your patience and guidance as a contractor will pay off in winning more jobs.

Once your customers know what they want, it is your job to see if their desires can be fulfilled. Will that oversized tub fit? Can a corner toilet be used? Is it possible to fit a double-bowl vanity in the existing space? Will his-and-her vanity areas work? (Fig. 1-10). Can the existing plumbing accept the additional fixture units of a new shower that will be used in addition to a replacement tub-shower combination? You will have to answer a lot of questions before you ever get beyond the paper stage of most jobs.

If you do kitchen remodeling, you must have heard of the work triangle. You don't have a work triangle in a bathroom, but there are similar design issues to keep in mind. For example, having storage available for daily toiletries is essential. A bathroom that has

Figure 1-10

His and her vanity areas are always a nice touch. <small>Courtesy of Ackermann</small>

mouthwash, toothpaste, and other daily routines sitting in plain view is not acceptable to many people. Even the location of the holder for toilet tissue is important. Towel racks and rings are other considerations. How high should the shower head be mounted above the floor of the fixture? It doesn't take long to develop a long list of questions. Most of the questions can be answered easily, but they do need to be addressed prior to commencing any physical work.

Any contractor is likely to assume that a toilet will be installed in a modern bathroom. But will the contractor give consideration to the type of toilet that will best serve the needs of the customer? Can a 1.6-gallon-flush function with old, cast-iron drains that are not graded properly? It can, but stoppages are likely. Maybe you should consider replacing the drainage piping or look for a vacuum-assisted flush. Would your customer be more comfortable using a higher toilet, such as a handicap-type toilet? Elderly people struggle more than young people do when getting on and off a toilet. The higher toilet can make life much easier for some people. This is a simple design issue, but it is one that is often overlooked.

What type of faucets will you install? If young children or elderly people will be using the bathing facilities, pressure-balanced and temperature-controlled faucets should be used. It only takes a matter of seconds for a person to receive serious burns from water that is too hot. Flushing a toilet while someone is in the shower can result in disaster if the right faucets are not used. What types of handles will the faucets be equipped with? Single-handle faucets are very popular, but there are still plenty of people who prefer the older, two-handle models. If your customer has arthritis, it would be wise to install blade-type handles, rather than round handles. The blade handles will be much easier for the customer to operate. These are all little things to most people, but they can make a big difference to some people. If you are on top of your profession, you will take the time to ask questions that will result in happy customers.

Lighting is important in almost any room, and bathrooms are no exception. The days of a single light over a metal medicine cabinet are gone. Fan-light combinations are very popular, but they are not enough when used alone. Consider putting a fan-light combination over the toilet and bathing area. Use strip lights on both sides of the vanity mirror. Cut in windows or skylights when you can for natural lighting (Fig 1-11). Since so many bathrooms are small, natural lighting goes a long way in making the rooms feel larger.

Will the bathing unit be equipped with a shower curtain or a glass enclosure? Most customers will prefer a glass enclosure, but there will be some who won't. Why wouldn't a person prefer glass doors over a curtain? Some people are afraid that their young children will be injured by the glass enclosure breaking. There are people who don't like cleaning soap scum off the doors and tracks. You won't know for sure what your customers want until you ask them, so don't be afraid to pose questions during your bidding visit.

Study your local code requirements. Either commit fixture-spacing rules to memory or create a simple chart that will tell you what your requirements are. For example, you must have a minimum of 15 inches of free space to either side of the center of a closet flange. If

Figure 1-11

Skylights are wonderful additions to bathrooms. Courtesy of
Andersen Windows, Inc.

you have an alcove that is 28 inches wide, you can't install a toilet in
it and meet current code requirements. You may also have to brush up
on your weight loads for floor joists. This is especially true if you are
installing an oversized bathtub or whirlpool tub. As you progress in
your experience, you will learn to expect certain questions. You will
also develop a sense of questions that you should ask to make your
designs more desirable.

It is a good idea to create checklists for yourself as your personal
needs become better known. For example, you could create a list that

asked key questions about flooring, wall coverings, plumbing fixtures, faucets, ventilation, electrical needs, and so forth. Such a checklist could be given to customers to complete at their convenience.

Specific design issues

Specific design issues may require you to seek advice from other professionals. For example, if a customer wants to have a shower installed in place of a bathtub, you could have a problem. Bathtubs can drain into a pipe that has a diameter of one-and-one-half inches. This is true even if the tub will have a shower head installed above it. Shower stalls, however, require a pipe with a minimum diameter of two inches. Failure to know this could cost you time, money, and frustration, not to mention the respect of your customers. If you are not familiar with codes pertaining to various trades, you need to either learn them or involve licensed tradespeople in your bidding process.

Architects should be able to answer any of your specific design issues. Be prepared, however, to pay handsomely for architectural advice. You may be able to find answers to questions in various code books. But if you don't have a solid, working knowledge of the trade involved, you may not understand the code properly. All codes are based on interpretation, and the wrong interpretation can get you in trouble. It is usually best to consult with an expert within the field of expertise that you have questions concerning.

Educate yourself

You should educate yourself so that you can be of more assistance to your customers. I've just told you to rely on the advice of professionals when searching for answers that are outside of your field of expertise. I stand by this, but I urge you to learn all that you can about every aspect of what you do. If you are a general contractor, you should have a general understanding of plumbing, heating, and electrical work. For example, you should know that a ground-fault-interrupter circuit is required in a bathroom. Knowing that a standard bathtub is five feet long is helpful. Being aware of what is involved in relocated ductwork or heating pipes is also important. The more you know yourself, the more you can share with your customers. There is another good reason for learning all you can about the trades you work with. By being on top of procedures for various aspects of a job, such as vinyl flooring, you can bid jobs more accurately and keep subcontractors a bit more honest by letting them know that you know a good deal about their trade. Some subcontractors will try to pull the wool over the eyes of general contractors. The only way to recognize and stop this is to know what the real deal is when an unscrupulous sub tries to stiff you.

Communication is the key

Communication is the key to success when you are working with people. This is true in almost any aspect of life, and it is certainly true in construction and remodeling. When you sit down with customers, you must remember that the people you are talking to do not possess your skills. If they did, they wouldn't require your services. Patience and strong communication are two major keys to making your business the best it can be. If you don't have good people skills, take steps to develop them. Read books, go to seminars, talk to consultants, do what you have to do, but make yourself effective in dealing with people.

Working with people is not always easy. It can be especially hard when you have two people making a joint decision and the couple can't come to terms with each other. You are the odd one out, so to speak. Basic sales skills will allow you to take control of the situation and focus it on the issues at hand. There are numerous books that teach sales techniques, and this type of reading should be a part of your ongoing education. Anyone who meets people and sells jobs can benefit from sales skills.

We don't have time here to turn this book into a sales lesson, but let me share one quick thought with you. Learn to stop talking and to start listening more. Many people talk so much that they never hear what their customers are saying. If you listen to people you will learn what they want. Don't clam up and say nothing, but give others a chance to contribute, and pay close attention to what they say. This is one of the most important steps you can take towards becoming a better salesperson and communicator.

The bottom line

The bottom line to finding the right plan is simple; give the customers what they want. Finding out what is wanted is the hard part, but you can do it. Don't rush your meetings with customers. Give them time and take time to make notes for yourself. Offer suggestions, but don't hog the show. Make sure that you and your customers agree on a plan of action before you offer pricing information. Once a plan is made, ask the customers to commit to it firmly. I have my customers sign each page of their plans and specifications as an exhibit attached to my work contracts. By doing this, I eliminate confusion and arguments before problems arise. You can do the same thing and make your life easier and your business more profitable. Now let's move onto the next chapter and talk about beautiful bathrooms and how they are made.

Beautiful bathroom designs

BEAUTIFUL bathroom designs don't just happen on their own. People like you have to make design plans that stand out from the mundane. I suggested in the last chapter that you refer to books of plans and magazines for layout ideas. Now I advise you to read every magazine available that deals with bathroom components and decorating. A stroll around your local bookstore will provide you with a wealth of resources from the magazine rack. Some magazines are published frequently, and others are special editions. I recommend that you buy all of them. The ideas you will get are well worth the price of the publications.

Beauty is usually thought of as a visual element. This being true, it's hard to beat the pictures that dot page after page in magazines catering to bathroom remodeling. There truly are a lot of magazines that feature and focus on bathroom design, decorating, and remodeling. I can't think of a better way to ignite your creative thoughts than paging from picture to picture. In fact, you can even start something of a scrapbook. Cut the pictures out of the magazines and compile a collection to show your prospective customers. If you don't like cutting up magazines, keep each issue and store them in magazine binders for review by your customers. Remember, ideas are paramount to the success of a great bathroom design.

Books also provide a lot of ideas for bathrooms. This book is only one example. If you look around, you can find many books that deal with decorating and other aspects of bathroom creations and remodeling. Some books concentrate exclusively on the finishing touches, and they are good additions to your work library. The more visual aids you can put together, the better prepared you will be to offer your customers the best services possible.

A mistake

Many contractors make a mistake when they plan a bathroom remodel. A lot of contractors think so intently on construction issues that they overlook the cosmetic aspects of a job. This is a mistake. Yes, the structural work is essential, but the cosmetic work is what most people see and appreciate. I'm not only talking about trim carpentry. No, all aspects of the finished job come into play. Tile, wallcoverings, light fixtures, bath accessories, and everything else that is open to public view control the overall effect a room has on an observer. Even color selections play a role in the success of a bathroom job (Fig. 2-1).

Figure 2-1

All aspects of the finished job come into play, including light fixtures and accessories. Courtesy of Wilsonart

▲
26

It may well be your job to concern yourself with just dimensions and spacing, along with all the other major construction elements, but don't overlook the fluffy stuff. Okay, so fluffy stuff may not seem like a lot to worry about, but it really is. Is a chrome faucet going to look as good as one in polished brass or antique brass? Will oak accessories look better than chrome units? Are towel bars better than towel rings? Would the walls benefit from a nice wallpaper border where they meet the ceiling? What effect would a skylight have on the bathroom? Could you make a strong statement by installing a beveled-edge mirror? Should a pedestal lavatory be used? (Fig. 2-2) Questions like these are just some of the ones that you should ask yourself before every job. Not all jobs justify going to extremes for outstanding results, but most do. Even if a job is a modest one, you should do all that you can to make it the best that it can be.

You may have to condition yourself to think in terms of finishing touches. Some contractors have a great deal of difficulty getting beyond the meat-and-potatoes part of a job. But believe me, the dessert is an important portion of what you are serving your customer. This is true not only for your existing customer, but for customers who will see the job and inquire about who did it. Much of the best business comes from word-of-mouth referrals. Having a host of jobs that showcase your special skills in the finished product are almost certain to get you more work.

Floor coverings

What types of floor coverings do you offer your potential customers? Nonabsorbent floors are normally required and used. There are times, however, when wood floors are installed in bathrooms. Personally, I try to dissuade customers from such a decision. Wood floors and water don't usually mix very well. Black spots are common when wood floors get wet and are not cared for properly. I can't recall any jobs that I've ever done where wood floors have been installed. However, I have seen wood floors in bathrooms and I have had customers request this type of flooring.

Figure 2-2

This pedestal lavatory has been installed with counter space on either side of it. Courtesy of Wilsonart

Sheet vinyl

Sheet vinyl is a typical flooring for bathrooms. It is the type of flooring that I have used most often. Vinyl is good for many reasons. It is affordable, easy to clean, water-resistant, and it wears well if it is of a good quality. One downside to vinyl is that it is slippery when wet. This can be a problem for anyone, but it is especially a concern for the young and the elderly. People who rely on crutches and walkers to get around are also at a higher risk when vinyl is installed. All in all, though, vinyl is an excellent choice for most jobs.

Tile

Tile has long been used as a flooring for bathrooms. It is considered an upgrade over vinyl. Sometimes too much tile is used (Figs. 2-3 & 2-4), but this is an exception rather than a rule. Being a premier floor

Figure 2-3

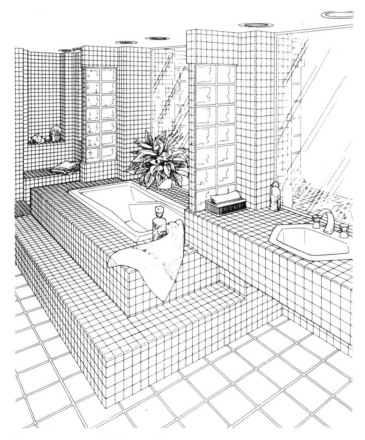

Excessive use of tile can be overpowering. Courtesy of Dal-Tile

Figure 2-4

Tile can be used on floors, walls, counters, and for showers. Courtesy of Dal-Tile

covering, tile enjoys great durability, but it has its problems. For example, the grouting can deteriorate to a point where it must be replaced. Grout joints can be very difficult to clean, and this is an irritant for some homeowners. Tile can be broken if a heavy object is dropped on it. Some types of tile become extremely slippery when wet. However, this can be overcome with the use of a textured tile. Price is another consideration when tile is to be used. The cost to install tile can be substantially more than what a vinyl floor would cost. Generally speaking, tile is a good-looking, durable, valuable floor covering that can be used in any type of bathroom (Figs. 2-5 & 2-6).

Tile for bathrooms come in many shapes, sizes, and styles. Most floors are done with ceramic tiles that are about four inches square. Quarry tile can be used, and mosaic tile is not uncommon. A wide variety of patterns can be created with tile. This adds considerably to your ability to create a unique bathroom floor. If you are looking for a type of bathroom flooring that will give you the ultimate in control for creativity, tile can't be beaten.

Baseboard trims

Baseboard trims for bathrooms are typically somewhat limited. Colonial trim is generally considered to be standard. Clamshell trim

Figure 2-5

Here is an example of a lot of tile being used in a tasteful way. Courtesy of Ackermann

Figure 2-6

Tile floors remain popular in bathrooms. Courtesy of Ackermann

is usually thought of as a cheap style of trim. There are, of course, other types of trim. Some types of homes use square-cut trim boards for baseboard trim. Bathrooms where tile floors are installed often implement the use of tile for a baseboard. Vinyl baseboards are, in my opinion, the least desirable type of trim for a bathroom in a residential home. Shoe molding is used frequently when vinyl flooring is used.

I strongly recommend its use. Some builders don't bother with the shoe molding, but this can lead to a curling of the vinyl as a floor ages.

Wall materials

Wall materials in bathrooms are typically water-resistant drywall. Other types of wall materials, like plaster, can be used, but drywall is far and away the product of choice. It is affordable, attractive, and durable under normal conditions. If flat walls that will be painted are the goal, I can't think of a better choice in wall materials. The least desirable wall material, in my opinion, is the type of wall covering that comes in sheets, usually 4'- × -8', that is supposed to look like tile or some other water-resistant design. At the very least, if this type of wall covering is to be used, it should be installed over drywall. My personal experience with the "water board" shows that it twists and warps easily and does not endure the test of time well.

Some bathrooms are of rustic designs. In these cases, weathered barn boards might be used as a finished wall surface. The moisture in a bathroom can cause problems with the wood, but this can be overcome, for the most part, with the proper staining or painting of the wood. The use of wood can enhance a bathroom, but it can also make the room appear very small and dark. I've seen bathrooms with Herringbone patterns of tongue-and-groove wood, clapboard style barn boards, and board-and-batten walls. My personal favorite was the Herringbone, but the weathered barn boards have their place in the proper setting.

Ceilings

Most ceilings in bathrooms are made of drywall. It may be painted as a flat ceiling or stippled and painted. Both types of ceilings are fine. I can remember a few bathrooms that were done with punched-tin ceilings, one or two that had tongue-and-groove wood ceilings, and too many that had dropped ceiling-tile ceilings. In most applications, I feel that painted drywall is the best choice for a bathroom ceiling.

You must keep moisture problems in mind when you are looking at ceiling choices.

Windows

Windows and skylights can make or break a bathroom (Fig. 2-7). Rooms without windows are often dull and confining. Not all

Figure 2-7

The windows beside this tub make the bathroom more enjoyable in a private setting. Courtesy of Hurd

Figure 2-8

Venting skylights help to control moisture in a bathroom, and they provide a lot of natural light. Courtesy of Vellux

bathrooms are positioned in homes where windows and skylights are possible. This is a shame. When you can use attractive windows to accent a bathroom, you have a definite advantage. Skylights add natural lighting and good looks. If the skylights are the type that can be opened (Fig. 2-8), you are providing an excellent source for reducing moisture build-up. Even if a light box has to be used for a skylight, it can be well worth the effort.

There are so many types of windows available to you that it would take entire catalogs to list them all. Every major window manufacturer produces windows for bathrooms. Some of the styles and designs are just out of this world. You can even get into stained glass, leaded glass, and etched glass if you really want to make your design special. Order window catalogs from your suppliers and look at all the options available to you. I think you will be surprised at how many choices you will have.

Doors

Doors are not usually a big part of a bathroom job. There is a door for the room and there might be a door or two for a linen closet. But doors just are not big elements in the style of an average bathroom. The door should normally be in keeping with other types of doors in the home. In regards to linen closets, you might have to choose between swinging doors and bi-fold doors. There could be a question of a flat door surface versus a louvered surface. On the whole, doors will not be a strong topic in your design stage. You might find that your customers prefer open shelves (Fig. 2-9).

Plumbing fixtures

The plumbing fixtures in a bathroom are generally key players in the success or failure of a bathroom's appearance. There is a smorgasbord of fixtures to choose from. Not only will you have an almost endless variety of fixtures to work with, the colors that are available are also numerous. There are standard colors, high-fashion colors, and colors that are fairly unusual. Toilets, bidets, lavatories, whirlpool tubs, shower stalls, standard bathtubs, and garden tubs are just some of what you and your customer will have to work with.

Toilets

A toilet is a toilet, right? Not really. Consider a toilet that has its flush tank hanging high above your head, connected to the toilet bowl by a flush tube and operated by a pull-chain. Does this sound like

Figure 2-9

Some people prefer open shelves to linen closets. Courtesy of
Wilsonart

some contraption out of the past? It should, because that is what it
is a replica of, and such toilets do exist. Have you ever seen a toilet
with a tank cover in the shape of a triangle. Well, corner toilets are
made this way. Review the sample of toilets I'm providing here.
(Figs. 2-10 & 2-11)

Figure 2-10

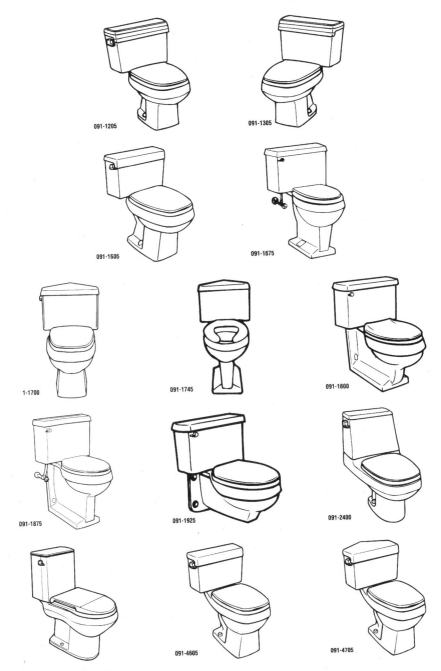

A variety of toilet types. Courtesy of Eljer

Figure 2-11

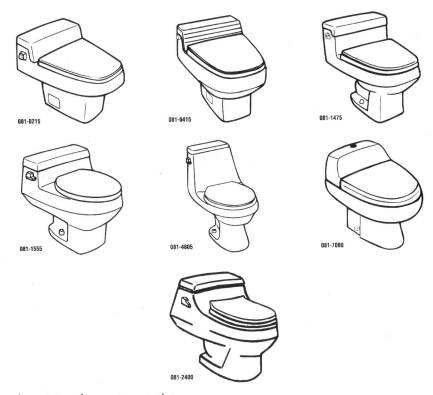

081-0215   081-0415   081-1475

081-1555   081-4805   081-7080

081-2400

A variety of one-piece toilets. Courtesy of Eljer

Bidets

Bidets are not common plumbing fixtures in the United States, but they are popular with people who want elite bathrooms. If your plumbers are not offering customers the option of bidets, you may be missing out on some business. It's true that bidets are not big sellers in most markets, but don't overlook this option when you are working with customers on a custom bathroom. (Fig. 2-12)

Lavatories

How many pedestal lavatories have you or your plumbers installed recently? Are you in a rut with vanities and cultured marble tops? Give yourself a break and look at other options. Wall-hung lavatories were

Figure 2-12

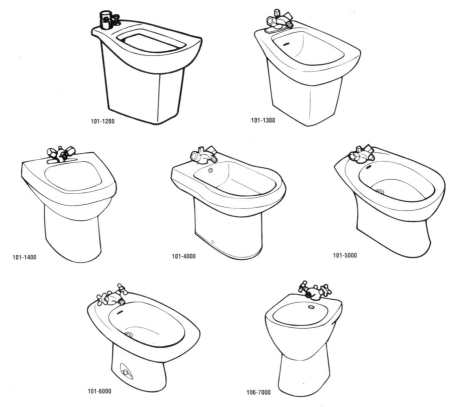

A collection of bidets. Courtesy of Eljer

the standard for many years. Vanities and tops took over the market and still dominate it. Pedestal lavatories don't offer any storage room, but they don't take up any more space than a wall-hung lavatory. Replacing a wall-hung lavatory with a pedestal lavatory is a major upgrade in style and appearance. Many pedestal lavatories are quite expensive, but there are builder-grade models available that are quite affordable. This is certainly a possibility that you should consider in your next job. (Figs. 2-13–2-16).

Tubs and showers

Bathtubs and showers have been in modern bathrooms for decades. This is not likely to change, but the types of bathing units you install for future customers could be very different from what has been the

Figure 2-13

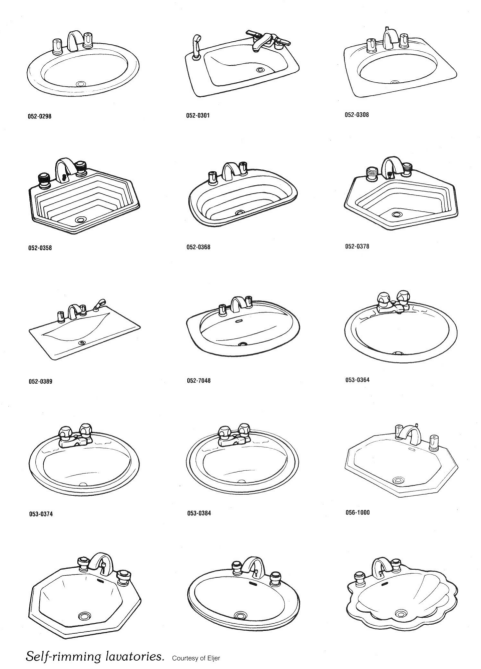

052-0298

052-0301

052-0308

052-0358

052-0368

052-0378

052-0389

052-7048

053-0364

053-0374

053-0384

056-1000

Self-rimming lavatories. Courtesy of Eljer

Figure 2-14

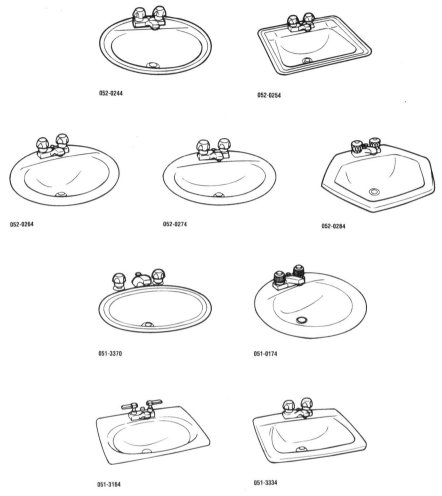

052-0244

052-0254

052-0264

052-0274

052-0284

051-3370

051-0174

051-3164

051-3334

Self-rimming lavatories. Courtesy of Eljer

standard over the years. When will this big change take effect? I don't know, but I expect that some entrepreneurs will come up with new bathing units that will be all the rage. For now, we have to work with what we have available to us.

Cast-iron bathtubs were very popular in luxury homes when I got into the trades. I remember this well, because the tubs weigh around 400 pounds and are a real pain to carry up stairs. Nowadays, my crews

Figure 2-15

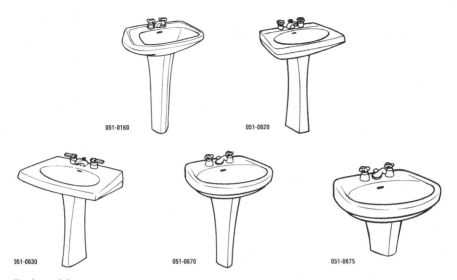

051-0160   051-0620

051-0630   051-0670   051-0675

Pedestal lavatories. Courtesy of Eljer

break up many more of these tubs than they install. When we demo a bathroom that has a cast-iron tub, we often take sledge hammers to the fixtures and break them into more manageable pieces. Most of the heavy tubs are replaced with fiberglass or acrylic tub-shower combinations. These types of fixtures are standards of the present industry. (Figs. 2-17 and 2-18).

Some customers still request cast-iron tubs. Why is this? There are a few reasons. The main reason might be that cast-iron has been noted for its superiority over the years. A more practical reason is that water from a shower head that is falling into a cast-iron tub does not make the pinging sound that it would in a steel tub. Generally, these are the

Figure 2-16

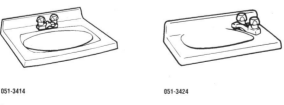

051-3414   051-3424

Lavatory tops. Courtesy of Eljer

Figure 2-17

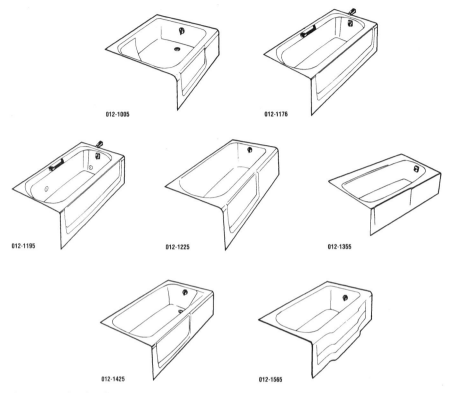

012-1005   012-1176

012-1195   012-1225   012-1355

012-1425   012-1565

Cast-iron bathtubs. Courtesy of Eljer

two reasons that I'm given for a preference towards cast-iron. There is no doubt that cast-iron tubs are tough, but their finish is not always as solid as the rest of the tub. If a heavy object is dropped in a cast-iron tub, it can chip the porcelain. This same type of problem can occur with steel tubs. Fiberglass units are much more durable when it come to hurting their finish. However, fiberglass can be broken quicker than cast iron.

Shower stalls can be chosen carefully to match a bathroom design (Fig. 2-19). They can be purchased with seats, one or two. Corner showers are nice in special circumstances. Showers with tops appeal to some customers. If you look around, you will find plenty of shower types to satisfy most any customer.

Figure 2-18

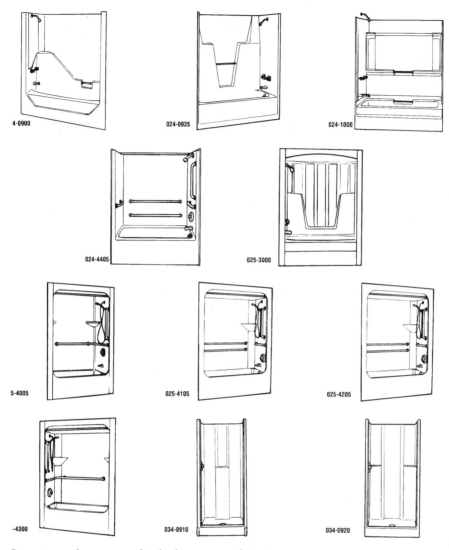

4-0900

024-0935

024-1000

024-4405

025-3000

5-4005

025-4105

025-4205

-4300

034-0910

034-0920

One-piece showers and tub-shower combinations. Courtesy of Eljer

Figure 2-19

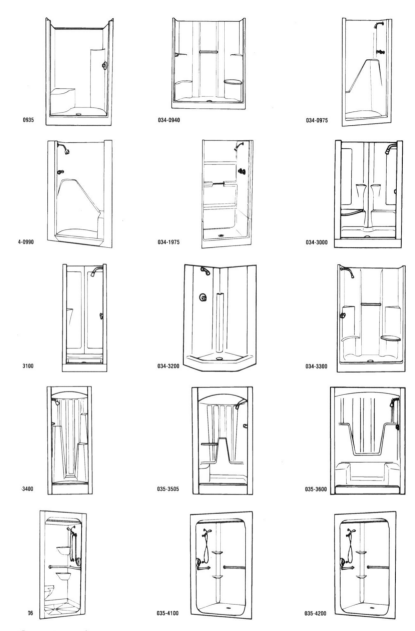

One-piece showers. Courtesy of Eljer

Whirlpool tubs

Whirlpool tubs are very popular. Small whirlpool units take up no more room than a standard bathtub. This type of whirlpool is not outrageously expensive and it is easy to install. However, most customers prefer larger, deeper tubs when space allows for such a fixture. Not all existing bathrooms offer enough room to install a two-, three-, or four-person tub. Added depth is usually easy to accommodate, but extended length and width can be a problem. You might be able to cantilever some additional space to make room for a wider tub. Another option might be to cheat some space out of an adjoining room, by moving a new bathroom wall into the available space. Deep tubs with whirlpool jets are relaxing and desirable, so look for ways to offer this option to your customers (Figs. 2-20–2-22).

Cabinets and closets

Cabinets and closets provide storage space in a bathroom. Most people crave all the storage that they can get. Many bathrooms depend on the use of vanities for storage space (Fig. 2-23). This is certainly not a bad idea, and there are a great number of styles and designs to choose from when shopping for a vanity. There are also some questions that you should ask your customer as you plan for a vanity.

How many doors would your customer like a vanity to have? Are drawers a preference? What quality is your customer looking for in a vanity. You can buy cheap vanities that look pretty good. However, the interior finish and the smoothness of the drawers can be lacking in low-end vanities. Study manufacturer fact sheets and learn all that you can about the products you plan to offer your customers. Get to know what combinations of doors and drawers are available in various sizes of vanities (Fig. 2-24). Invest some time in reading up on quality construction and how it will affect your customers. Product knowledge is very important when it comes to making sales.

Linen closets are sometimes found in bathrooms and sometimes they are in a hall, near a bathroom. Finished cabinetry is available in the

Figure 2-20

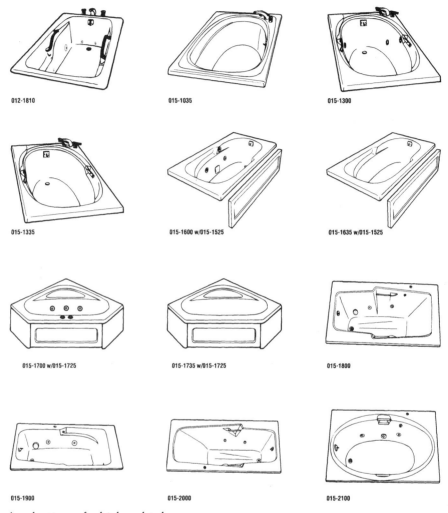

A selection of whirlpool tubs. Courtesy of Eljer

form of linen storage. The advantage to buying a cabinet is that you or your carpenters don't have to build a closet. Another is having the linen area match the vanity in style and color. Stand-alone cabinets are a good choice when it comes to linen storage.

Medicine cabinets are not standard equipment in bathrooms, but they are handy storage areas. You can choose between flush-mount and recessed units (Fig. 2-25). If you are planning to install a recessed cabinet where one has never been before, you must make sure that there will not be vent pipes or ductwork in your way. Both types of cabinets have their own appeal. A selection will be up to your customer, but it's easy to offer a wide array of medicine cabinets for your customers to choose from.

Figure 2-21

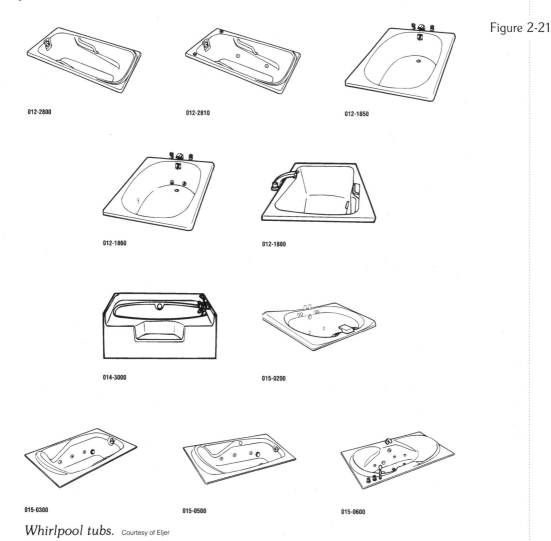

012-2800   012-2810   012-1850

012-1860   012-1880

014-3000   015-0200

015-0300   015-0500   015-0600

Whirlpool tubs. Courtesy of Eljer

Figure 2-22

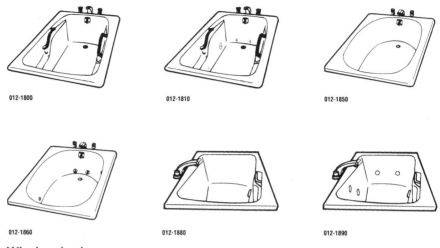

012-1800   012-1810   012-1850

012-1860   012-1880   012-1890

Whirlpool tubs. Courtesy of Eljer

Figure 2-23

A bathroom design with vanity and wall cabinets. Courtesy of Universal Rundel

Figure 2-24

Vanities that combine drawers and doors are very functional.
Courtesy of Universal Rundel

Figure 2-25

*Here is a bathroom with a flush-mount medicine cabinet and a wall
cabinet.* Courtesy of Universal Rundel

Light fixtures

Light fixtures are essential elements in a bathroom. Many rooms do
fine with simple fixtures, but some bathrooms cry out for unusual
lighting. You might use track lighting, recessed lighting, strip lighting
(Fig. 2-26), or traditional fixtures. One key is to provide plenty of light
sources so that the bathroom will not be dark and appear small. Far
too many bathrooms are equipped only with minimal lighting, and

this is a big mistake. Yes, it costs more for extra lighting, but the expense is worthwhile.

Visit local lighting stores and obtain catalogs of what they have to offer. Your electrician will probably be happy to provide you with catalogs. As a contractor, you benefit greatly from building a reference library that potential customers can use and choose from. Most contractors offer their customers a lighting allowance for the purchase of light fixtures. This is a good practice, but make sure you budget enough into the job to allow your customers a wide selection of what is available.

Figure 2-26

Strip lights, like those used here, are very popular. Courtesy of Universal Rundel

Figure 2-27

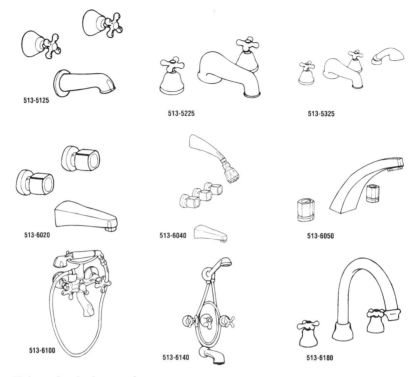

Tub and tub-shower faucets. Courtesy of Eljer

Faucets

Faucets will be needed for bathing units and lavatories. This should come as no surprise, but you may be shocked to see how many options there are for your customers to choose from. We will talk about faucets more when we get into the plumbing chapters, but take a moment now to make yourself familiar with the abundant quantity of styles available when it comes to faucets. (Figs. 2-27–2-35).

Hardware

Hardware for vanities, closets, windows, and doors is often taken for granted. Many contractors give their customers whatever hardware comes on prefab units (Fig. 2-36). This is okay, but you can make some decorating statements of your own by replacing stock or

Figure 2-28

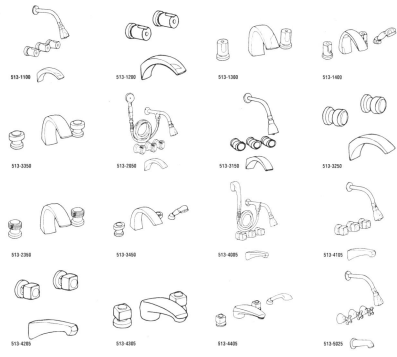

Tub, tub-and-shower, and whirlpool faucets and fillers. Courtesy of Eljer

Figure 2-29

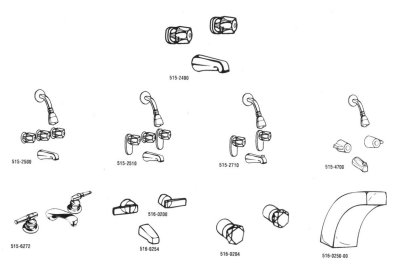

Tub, tub-and-shower, and whirlpool faucets and fillers. Courtesy of Eljer

Figure 2-30

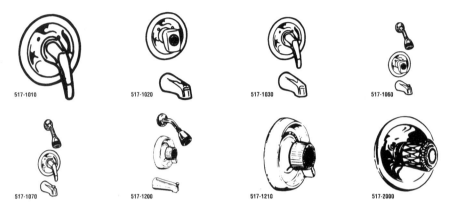

Single-handle tub and tub-and-shower faucets. Courtesy of Eljer

standard hardware with something more exotic. A number of companies specialize in hardware, so there is no excuse for accepting stock material as an only choice. Replacing existing hardware with designer models can turn an ordinary bathroom into a showplace. Remember that much of your future work may come from jobs that you have done and that other people have seen. By creating bathrooms that will be talked about, your company will be talked about and probably called upon for new jobs.

Accessories and accents

Accessories and accents are part of the finishing touches needed to make a bathroom beautiful. You can accent with colors (Fig. 2-37). Mirrors, roll-top doors (Fig. 2-38), and wall shelves can add to the look of a bathroom. Your customer may choose standard chrome materials for basic accents, such as towel bars, but oak or polished brass could make a big difference in the overall effect the room has on a person. Remember to key electrical outlet and switch covers to the type of accessories used. It's silly to invest in rich wood accessories and use cheap, plastic, ivory-colored electrical covers. Pay strong attention to detail and see to it that all elements of the bathroom fit well together.

There is no magic wand that you can wave to make an outstanding bathroom, but there is a lot of creative work that you can do to

Figure 2-31

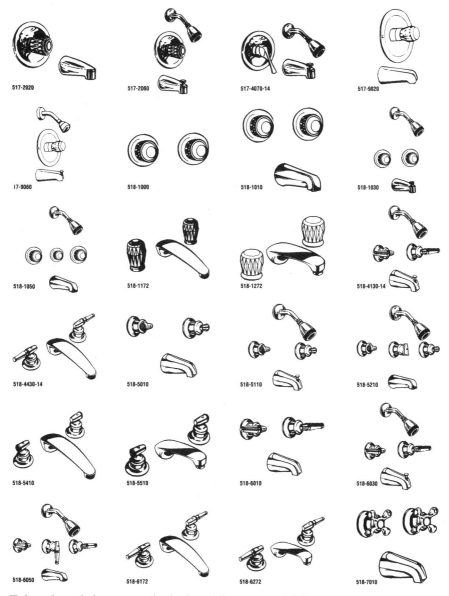

517-2020
517-2060
517-4070-14
517-9020

17-9060
518-1000
518-1010
518-1030

518-1050
518-1172
518-1272
518-4130-14

518-4430-14
518-5010
518-5110
518-5210

518-5410
518-5510
518-6010
518-6030

518-6050
518-6172
518-6272
518-7010

Tub, tub-and-shower, and whirlpool faucets and fillers. Courtesy of Eljer

Figure 2-32

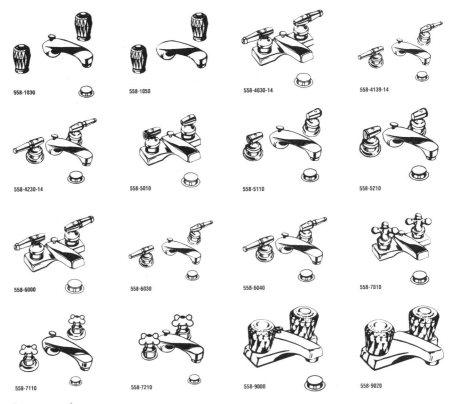

Lavatory faucets. Courtesy of Eljer

achieve your goal. Customers often don't have the experience or the vision to tell you what they want or what will work best. It's your job, as the contractor, to step in and advise the customers. Before you can do this, you must be knowledgeable. You've taken an excellent step in reading this book, but read all that you can on the subject of bathrooms. There is no shortage of reading materials, ideas, or special construction and decorating options. The only reason I can think of for a contractor to fail at building and remodeling award-winning bathrooms is a lack of knowledge or interest. If you want to become an elite contractor within your field, you can.

Figure 2-33

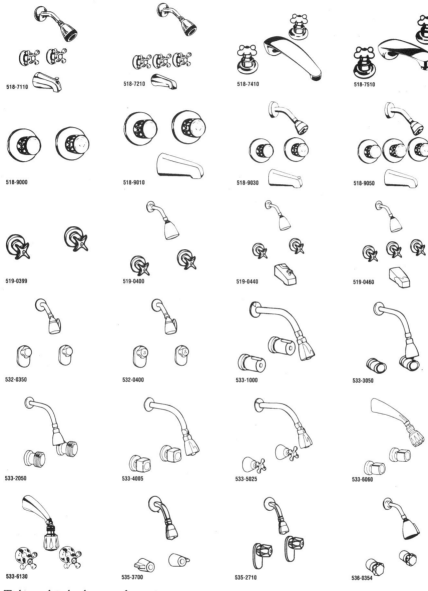

518-7110   518-7210   518-7410   518-7510

518-9000   518-9010   518-9030   518-9050

519-0399   519-0400   519-0440   519-0460

532-0350   532-0400   533-1000   533-3050

533-2050   533-4005   533-5025   533-6060

533-6130   535-3700   535-2710   536-0354

Tub and tub-shower faucets.   Courtesy of Eljer

Figure 2-34

552-1701 · 553-1000 · 553-1105 · 553-1300 · 553-2350 · 553-2050 · 553-2250 · 553-4005 · 553-4165 · 553-5125 w/ HANDSPRAY · 553-6000 · 553-6020 · 553-6040 · 553-6060 · 553-6100 · 553-6140 · 553-6180 · 553-6190

Lavatory faucets. Courtesy of Eljer

Figure 2-35

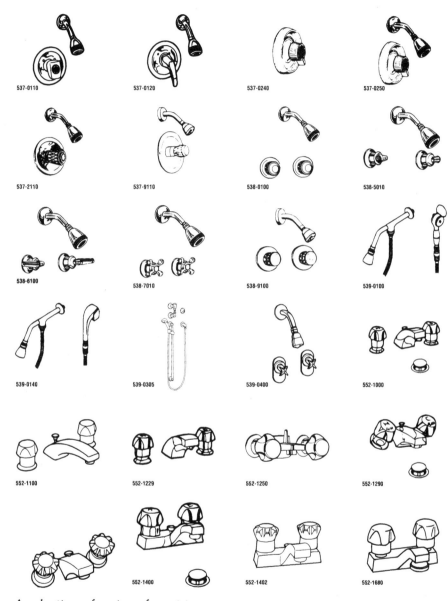

A selection of various faucet types. Courtesy of Eljer

Figure 2-36

Stock hardware is fine, but replacing it with custom hardware can be advantageous. Courtesy of Wilsonart

Figure 2-37

This bathroom has been accented with colors. Courtesy of Wilsonart

Figure 2-38

Here is a bathroom with multiple accents. Courtesy of Wilsonart

Cost estimates

COST estimates are a cornerstone of any contracting business. If they are done correctly, you win jobs and make money. When your estimates miss the mark on the high side, you lose jobs. If the prices you come up with are too low, you'll have plenty of work and very little money when the jobs are done. You have to learn to create accurate cost estimates that you can depend on. Estimating the cost of remodeling work is much more difficult than pricing new construction. There are so many things that can go wrong with remodeling jobs. To be safe, you need to build in a buffer of money to cover unexpected costs. However, if you protect yourself too well with an inflated bid, you will not win many jobs. This means that you have to learn how to anticipate problems before they occur. Doing this after many years of experience in remodeling is not difficult, but doing it when you are first getting into remodeling can be a major problem.

Remodelers who are rookies to the business don't have enough experience in their field to see the signs of a cost overrun in the making. Trained eyes can spot rotted floors, moisture problems, and other elements of a job that can run the costs up. Making mistakes with bids in the early stages of a career is to be expected, but it's a costly learning experience. If you can gain the knowledge of what to look for without making the mistakes, your bank account will grow much faster.

The best way to prepare yourself for accurate cost estimates is to immerse yourself in a lot of remodeling work. One way of doing this is to work as an employee for a remodeling company until you have several years of hands-on experience under your belt. Another approach is to take the earn-while-you-learn approach, as I did, and make your mistakes and gain your knowledge as you go along. A third way is to read books, like this one, talk to experienced contractors, and take as many subcontractors as you can with you on estimates. Most contractors find themselves using some or all of these methods.

There is a common mistake made among business owners who don't have a lot of field experience. The owners think that they can buy their estimating ability with computer software and printed cost guides. We will talk about these two approaches to estimating soon, but let me just say now that you cannot buy estimating ability without hiring an experienced estimator to work for your company. There simply is no substitute for experience.

Estimating a job involves many factors. You will have to figure the cost of materials, an expense for labor, a figure for overhead expenses your profit, and your other costs of doing a job. Some of the costs are considered hard costs and others are soft costs. An example of a hard cost could be a bundle of lumber or a bathtub. Soft costs are made up of less obvious elements, such as building permits and blueprints. All costs must be taken into consideration when you are putting together an accurate bid. Leaving out overhead expenses, such as insurance and advertising will result in weaker profits by the end of the year. Figuring the cost of a job is much more complex than many contractors believe it to be. For you to be successful and profitable you must not take this important part of your work lightly.

Overhead expenses

Overhead expenses are costs that you incur while you are searching for work, doing jobs, and waiting for your next job to come along. In other words, overhead expenses are continuous and don't go away unless you eliminate them altogether. Expenses for overhead can drag a company down in just a few months. If you don't have any work coming in but you have operating costs going on, you are losing money. It can take a big profit from the first few jobs you do to get yourself even with the money you have invested. You can't expect to build a business without having overhead expenses, but you must budget for the costs and recover them from each job you do. It's unreasonable to try to get all of your overhead return from one or two jobs. Most contractors establish a percentage to use when factoring their overhead expenses in a job bid.

How can you know how much overhead you should cover with each job? First you must do some budget forecasting. You have to know approximately what your overhead expenses will be for the entire year. Once you know this, and it's easy to figure out, you can establish a percentage to tag onto each job. Okay, I said it was easy to figure out your expenses, so let me show you just how simple it is. You will need paper and a writing utensil to record your budget.

Start your overhead budgeting by making a list of all your everyday expenses that will be incurred on an annual basis. For example, put in a figure for your rent, utilities, telephone expenses, advertising, and so forth. Many of the numbers will be known expenses. Your rent shouldn't go up if you have a lease, so you can safely project what the cost of having an office for the year will be. Utility bills fluctuate. But you can either make a best guess or you can check with your landlord to see what previous utility costs have been. Most landlords who deal in commercial properties have such records available. The basic telephone service will be constant, and you should be able to estimate a reasonable figure for long-distance charges. If you will be responsible for paying someone to cut grass or plow snow as a part of your business expenses, you have to include these costs in your budget.

What else do you have to pay for all the time? If you have an ad in the local phone directory you should include its cost in your budget projections. Will you advertise on television, radio, or in newspapers? What plans do you have for a direct-mail advertising campaign. Establish an intended budget for your advertising that will cover the entire year. Make your allowance a little on the high side. It's better to have your projects running higher on paper than what they actually will be than to figure them too cheaply.

Insurance is an ongoing expense. Business and trade licenses cost money whether you are working or not. If you have a pager or a cell phone, you should include the estimated costs for them in your budget. The base cost will be known and you can project added expenses for calls made. Will you have a human answering service that must be paid monthly? If so, put the cost in your spreadsheet of figures. Are we done yet? No, not likely. What about the cost for your trucks and the maintenance required to run them and to keep them on the road? If you miss expenses in the budgeting stage your profits will suffer by the end of the year. Take the time early to make sure that you are including all costs. Did you remember to figure in the costs for self-employment tax, worker's compensation, payroll, and payroll tax deductions? Not all businesses run the same and their expenses vary both in type and amount. Computing your overhead expense is something you must do on an individual basis.

Now let's assume that you can complete your comprehensive listing of all overhead expenses. They are probably higher than what you would have guessed, but at least you have realistic numbers to work with. Now you have to figure out a way to get some of the expense money recovered fairly from each job you do. This part of the equation is a little trickier than jotting down overhead numbers.

To establish a percentage of overhead to charge off to jobs you need to know approximately what your annual work volume will be. For example, a small contracting company might gross only $120,000 for a year. A larger company might make $500,000. Builders can gross well into the millions of dollars. Before we go on, let's discuss gross money, net money, and what I call net-gross money. Gross money is all the money that you receive during the year. Net money is the amount of money that is left over after all expenses have been paid. Net-gross money is the money received from a job after overhead expenses, hard costs, and soft costs have been extracted, but there are still some expenses to come out of the profit. Net money is the key amount, because that is what you will have to spend or save when your year is done.

Now let's say that you plan to gross about $200,000 this year. When you total your overhead expenses, you see that they amount to $2,500 a month or $30,000 a year. If you divide your overhead by your gross ($30,000 divided by $200,000) you will see that the overhead cost is 15 percent of your expected gross. This tells you two things. First of all, you will have to factor in 15 percent on top of your true costs for a job just to cover your overhead expenses. Such a large amount is too much in most cases. By the time you add a percentage for profit you will probably be pricing yourself out of the market. This means that you have to either do more business or reduce your overhead. What would the numbers look like if you could do $500,000 in sales? Your overhead percentage would drop from 15 to 6, and this is much more realistic, though still a little on the high side. There is no set rule on what percentage of your gross income your overhead must be. You must determine what you are comfortable with. A 10-percent figure may work out just fine for you, but the lower you can keep the overhead without affecting business volume and quality, the better off you will be.

Some small contractors don't factor in a salary figure for themselves. They consider the annual profit of the company to be their salary. If this is the case, overhead numbers change from the example I gave above. The earlier example included a modest salary. If you figure your income on profit rather than on a weekly paycheck, your overhead expenses could be much less. For example, if you work out of your home and don't pay yourself a salary in addition to your profits, your overhead might be around $500 a month, or even less. If you sell $200,000 worth of work this way, your overhead percentage is only 3 percent, which is pretty darn good.

Now that you can see how percentages are arrived at for overhead, let's put the knowledge to work. Assume that you are bidding a bathroom remodeling job and your first figures are showing a bid price of $9,000. This figure represents all expenses, except for your overhead. If you are working on a mark-up percentage of 5 percent for overhead, the total bid price would be $9,473.68. The extra $473.68 that is computed into the job estimate is what will allow you to recover your cost of being in business. If you end your year with at least as much gross income as you projected and you keep your expenses to a level that was projected, you will have a healthy business.

Preliminary expenses

Preliminary expenses are sometimes overlooked when remodelers are figuring job costs. What is a preliminary expense? It can be any number of things. For example, your time while working on bids is an overhead expense. We just finished talking about these. But paying for a set of plans to be drawn for a job might be a preliminary expense. What do I mean by a preliminary expense? I'm talking about money that your company will have to spend before physical work is started on a job. It could be permit acquisition, arranging for a mobile trash container to be put on site, or surveying fees for an addition being built.

Preliminary expenses are not usually huge fees, but they do add up and they can be expensive. When you estimate a job, you must cover all costs, and I do mean all costs, in your estimate. Not thinking about the need to build a trash chute from an upstairs bathroom window will not cost you a fortune, but it will eat into your profits. Failure to think about needing dust containment is not a big deal, but it's more money down the drain for you. Sit down and study each job before you submit a bid. Think the remodeling or building process through from start to finish, with the start being established as the moment your contract is signed. List every needed activity and account for it in your cost figures. This is the only way to make your estimates accurate and profitable.

Dissecting your bid phase by phase

Most experienced remodelers and builders do okay when they are pricing their jobs on a phase-by-phase basis. They may make mistakes in their figures, but at least they make some attempt to account for them. The overhead expenses, preliminary expenses, and final expenses are the categories that come back to haunt most contractors. I believe the reason for this is that most contractors are tradespeople first and business owners second. There are an awful lot of excellent tradespeople in the field who are not making the kind of money that they should as independent contractors. The reason for this is that they are so consumed in doing their field work that they don't maintain a level of administrative interest that is needed for a business to run at maximum profits.

When I started into business for myself I was the only person in the business. I tried to maintain a businesslike approach and did pretty well, but still I let some things slip through the cracks. As my company grew and I hired employees, I had to take more of a business approach and less of a trades approach to what I was doing. By the time I was grossing $4 million a year, I had a heavy schedule of business involvement and very little field work on my hands. At one point, I had over 120 people answering to me, and there was plenty of administrative work to be done. I had a competent office staff of three people and two field superintendents, but my need for attention to business elements was still extreme. After playing the contracting game both ways, I can appreciate the one-person company and the moderate corporation. My experience with major corporations has come has an employee, so I do not have the personal hurt or excitement from that situation, but I do know how it's done in the big leagues.

When you go out to inspect a job for a bid, you have to look at more than just the obvious, physical work. How are you going to deal with the debris from demolition? Is there a secure place to store equipment or materials on the job site? What preparation will be required to protect the customer's home from the traffic and work created by your crews within the home? All of this type of work has a price tag, and you have to account for it. If you don't, the unexpected costs will come out of your profits.

In addition to your observation of planning stages, you must look for existing conditions that might affect the job. For example, is there adequate room in door openings and along halls and stairways to get fixtures and other materials into the bathroom? It would be pretty embarrassing to sell a job and then find out that the tub-shower combination or the whirlpool tub that you contracted to install will not fit in the house. A situation like this would hurt your reputation and put a serious dent in your profits. Look around when you inspect a job. Pay close attention to everything and make notes. Your notes can be very helpful when calculating a price and when scheduling subcontractors. To put things in perspective, let's do a little phase-by-phase dissection of our own.

Start-up

What is required before the start-up of a job? Some type of container for trash removal will be needed. You might use a truck or a mobile trash container. Are you going to build a trash chute from the bathroom window, or will your crews carry out the demolition debris? Will you have to cover carpets or wood floors to protect them as you make a path to the bathroom? Will the homeowners have to remove personal possessions from shelves or walls to keep them from being broken as your crews pound away with the remodeling? If you are building an addition you might need a survey done. Can you get a concrete truck to a good position to pour footings for an addition? Have you discussed lawn damage with your customers? Omitting steps like these from your bid process can be costly. It can also hurt your standing with customers. Cover as many bases as you can before you go to contract with a customer.

Demolition

Demolition is a common part of remodeling. This is true of a simple remodeling job or of an extensive addition. If you are building an addition, you might not think of demolition work. But you've got to attach the addition to the existing house in some fashion, and this will create the need for some amount of demolition work. When you are doing a remodel on an existing bathroom, you will probably be removing fixtures, flooring, and maybe more. This is not usually a large job, but it is a phase you have to plan for. How will your customers get by during the time that the bathroom is out of order? Is there another bathroom in the house? Will you have to supply a portable toilet on the job? How likely are your crews to damage some other part of the home as they are tearing out the bathroom? If you expect that nail-pops will occur in a ceiling below the bathroom from all the banging and pounding, you had better address the issue in your bid.

Existing conditions

Existing conditions can skew your bid prices considerably if your contract doesn't protect you properly. For example, if you think that you are going to pull up an existing vinyl floor and replace it with a new one you would give one price. But, if you found that underlayment was going to be needed, and it often is, your price would be higher. Evidence of a rotted subfloor around the base of a toilet would increase the price even more. A little probing with a strong knife or screwdriver can give you a good indication if rot is present around a toilet. Knowing in advance what is going to be required to wind up with a good-looking finished product is important.

Does the existing bathroom have a ground-fault-interceptor circuit? Code requires one. How difficult will it be to isolate the circuit? Will you simply replace the existing outlets? This is probably the most sensible solution. How far will you have to strip the walls to get a good finished product? Can you get by with some touch-up and stain blocker, or will you have to replace existing wall materials before applying a finished covering? Does the bathroom have adequate ventilation to meet current code requirements? Will existing floor joists support the added load of an oversized bathtub? There are a multitude of potential problems waiting for you in a remodeling job. Let's look at some of what you might run into on a phase-by-phase plan.

Flooring

Flooring is one area of bathroom remodeling where a lot can go wrong, but the trouble can be seen in advance, most of the time, if you know what to look for. Does the existing floor slope? Is there any rot around the base of the toilet? Can you see telltale signs of cracks showing through existing vinyl? Are existing tiles buckled up? Can you bounce on the floor without having too much shake in the room? How old is the house? Do you suspect multiple layers of existing flooring? Can you make a threshold transition easily? Start looking for signs such as these to see if you are getting into more than a normal replacement job.

You could simply plan on installing all new subflooring, underlayment, and finished flooring, but this will probably price you out of the competition. It's possible that you could install a new vinyl floor over the existing one without any removal. Do you have the skill to tell if this is feasible? If you don't, ask your flooring contractor to look at the job with you. Most subcontractors are willing to go to job sites before making a bid. In fact, many subs prefer seeing a job before bidding it. Homeowners may not like a parade of subcontractors coming to their homes, but there is a good way for you to defuse this complaint and turn a customer's concern into appreciation for your company. Let me explain what I mean.

I've had customers who were displeased that I wanted to bring a number of contractors into their homes prior to giving a firm price on a remodeling job. Most of these jobs were much larger than a simple bathroom deal, but the principals I used should work for you, too. Explain to your customers that in order to give them the best price possible, your subcontractors need to see the job. Let the customer know that contractors or subcontractors who must give a price on a job that has not been inspected are forced to keep prices higher to allow for existing conditions that are not known. Once you let your customers know that the slight inconvenience will likely result in a lower price, you will probably achieve their approval for various inspections.

Having individual subcontractors look at a job before offering estimates is good protection for you. If your plumbers give you firm prices, you should be able to count on them. The same goes for your electricians, flooring contractors, and so forth. If you take it upon yourself to estimate all costs on your own, without help from your subs, you are much more likely to lose money on jobs.

Walls and ceilings

Walls and ceilings don't usually hold many surprises. This is a phase of the job that most any experienced contractor can deal with individually. However, you might notice stains or moisture conditions that would support the consultation of experts, such as painters. If you are a pure business owner with limited to no hands-on experience in the trades, you might do well to talk to drywall contractors and other contractors about the condition of existing walls and ceilings.

Plumbing

Plumbing is a significant part of bathroom remodeling, and plumbers usually have steep hourly rates. This, certainly, is an area of work that you don't want to miscalculate. If you can tell that the only work required is a straightforward fixture replacement, there is probably no need to call your plumber in for a look-see. However, if fixture locations will be changed or there will be substantial changes in the types of fixtures installed, asking your plumber to take a look at the job is a good idea.

There are a number of behind-the-scene situations that can run up the plumbing cost on a bathroom job. For example, the pipe under the toilet may not have a flange. You could find that the pipe is made of lead and cannot be made to work with modern materials. This could cause you to tear up part of the floor or part of a ceiling below. Water pipes might be galvanized steel or brass pipe, and this can run up costs. Proper traps may not be present for new fixtures. If you don't know plumbing and remodeling well, ask your plumber to accompany you on your estimate visit.

Electrical work

Electrical work does not usually amount to much in a bathroom job. Adding ground-fault-interceptor (GFI) outlets is common. Replacing existing light fixtures is normal, but not usually complicated. Unless you will be running new wiring for fixtures or devices that have not been present in the existing room, you probably don't have to worry about calling in your electrician to see the job. One exception to this would be an old house, where existing wiring is substandard to begin with.

Heating

The heating systems in bathrooms can run the gamut of wall heaters to radiators. Depending upon what is present in an existing bathroom, you might need some help in this area. Modifying hot-water baseboard heat is usually simple. Electric baseboard heat can be removed, replaced, or relocated without a lot of problems. Wall heaters don't normally offer a contractor much trouble. Radiators, however, can be a major pain in the neck. If they are moved, they are likely to leak when they are reinstalled. Taking sections out of an existing radiator or adding new sections can be very tricky. Ductwork, for central heating or air conditioning can also be troublesome, due to its bulk. Fortunately, it is rarely necessary to do much with the heating system in a simple remodeling job.

Drywall, paint, and wallcoverings

Drywall, paint, and wallcoverings don't normally produce bad surprises. If you are concerned about existing conditions accepting new wallcoverings, don't hesitate to talk it over with your subcontractors. Typically, the walls and ceilings will not pose a problem for you. Check the walls to make sure that they are square and plumb. Inspect for stains or moisture problems that might prohibit your proposed plans. In most cases, you will find the conditions to be normal and acceptable.

Tile work

Tile work, whether existing or new, can cause some trouble if the conditions are not right. If you are putting tile on a floor where there has never been tile, you might have to beef up the support or the subfloor. Tile around a tub or shower might require you to replace the drywall behind existing tile or wallcoverings. Since adhesives are used in most tile applications, the surface being tiled must be clean. This sometimes means tearing out existing wall materials and replacing them with water-resistant drywall.

Doors

Doors rarely offer many problems. Check the door openings to see that they are plumb. Confirm dimensions to make sure that standard doors will fit the existing openings. So long as the openings are a standard size, large enough to allow you to get new fixtures in the room, and plumb, you should not have any trouble. Still, this is something that you should check and note before giving a firm price on a job.

Windows and skylights

Windows and skylights can be an expensive part of a bathroom job. You might be leaving existing materials in place or you might be replacing them with something new. If you will be doing replacements, check the existing openings to see that they will accept sizes available to you. In the case of new skylight installations, look to see if you will have to build light boxes, and if so, how large they will have to be. Confirm access to the attic for skylights. Not all homes have scuttle holes or pull-down stairs. When existing windows will be left you should check their condition. A window that looks okay could have rotted wood all around it. Moisture is often a problem in bathrooms, and windows can be one of the first places that the moisture attacks.

Take your time

One key to success in working up estimates is to take your time and be sure not to leave any expense out of your figures. There are, of course, other issues at hand. You have to work with subcontractors and suppliers to get prices, for your estimates. We will talk more about this in the next two chapters. Good organizational skills, a watchful eye, and patience are all key elements of accurate bids. If your prices are off, your profits will be off. Either you will have little to no work because of high prices or you will be swamped with low-paying work because you underbid the job. Seriously, take your time and take notes when you do your estimate inspections. Figure your bids when you will not be distracted. Once you have an accurate number to work with, try to stick to it. Negotiating with customers is necessary and sometimes you have to take work for lower prices to survive, but do your best to ensure a good profit from every job.

Subcontractor bids

IF YOU are a general contractor, subcontractor bids are the heart of your business. Without your subs, you don't have the personnel to get jobs done. This creates an interesting business situation for you. On one hand, you are in control as the general contractor. Subcontractors should be flocking to you for work. But, on the other hand, you are dependent on subcontractors to make your living. This can be a delicate balance. You need to be in control, but you also need to keep your subcontractors happy. Walking this line can be difficult in the best of times, and it can be really tough when you are starting out or having financial troubles.

Some remodelers don't maintain any payroll crews. Depending upon every trade to be a subcontractor has financial advantages, but there are also risks associated with this type of business plan. Many business owners have in-house crews who are capable in certain roles, such as carpentry. In a case like this, subcontractors are limited to specific trades, such as plumbing, heating, and electrical work. It's fairly unusual to find a general contractor who employs licensed trades for all forms of work when a trade license is required. There are, however, some companies who have in-house, licensed tradespeople to do their work. For a company to afford to have all aspects of bathroom remodeling contained with in-house employees, the company must have a large volume of work to do.

How is your company set up? Do you rely heavily on subcontractors? Do you use employees for most of your work? How is your profit picture at this point? Subcontractors can appear to cost more than employees, but this is not always the case. Sure, they cost more on an hourly or per-job basis, but when all cost associated with both types of workers are assessed, subcontractors can be a bargain. Before we start into the meat of this chapter, let's discuss briefly the pros and cons of subcontractors when compared to hourly employees.

Subcontractors or hourly employees?

Which is a better bargain for you, subcontractors or hourly employees? It might be hard to answer this question off the top of your head. If you've been in business for a number of years, the answer might jump out at you, but inexperienced business owners often have not run the numbers both ways. Are you better off to employ someone who can install tile or should you sub the work out? It depends on how much tile work you have. To illustrate this, let's look at a quick comparison.

Assume that you hire someone who can install tile for you. This person is going to cost you some set hourly fee. In addition to the hourly wage, you will have to pay payroll taxes. There will probably be paid holidays and some paid vacation for you to pick up the tab for. Health insurance is another potential benefit that you may pay for. Do you supply a truck and auto insurance for the installer? What will your tile installer do when there is no tile to install? A lot of employees wind up killing time, doing inventory work, washing trucks, or making some other kind of work to do when jobs are slow coming in. If you are paying for this time, you are probably not making any money for the time being paid for. My point is this—employees cost much more than their hourly wage.

If you compare a subcontractor's hourly wage to that of an employee, you will probably see that the subcontractor costs substantially more. This fact can be more of an illusion than a reality. When you retain subcontractors, you don't have any of the other costs to pay. This means that the hourly rate is the whole deal. To get a fair comparison, you have to add up all of your employee cost and compare it to the expense of the subcontractor. Unless you have a steady flow of work for your employees, subcontractors will frequently prove to be a more cost-effective approach to solving your labor needs.

I've known many general contractors who tried to save money by hiring employees and eliminating subcontractors. In most cases, their plans didn't work. You really have to maintain an abundance of work to justify hourly employees in the field. Not only do you have initial costs to consider, you have to think about call-backs. If you send your in-house plumber on a job and something goes wrong after the job is completed but still under warranty, you have to pay the plumber to go back at your expense. This is not the case with a subcontractor. When a subcontractor's work fails, the sub must go back, without additional pay from you or the customer, to correct the deficiency. This can be a definite advantage to you.

My experience has shown that subcontractors are usually the best way to go. I like to keep a few hourly people on board to take care of problems when I can't reach a sub quickly, but I generally rely on subcontractors for my production work. After years of trying it both ways, I definitely feel that subcontractors are the best overall way to structure a contracting business. There are, of course, exceptions and you will have to weigh your own personal needs to determine which type of worker will prove best for your company.

Soliciting bids

Soliciting bids from subcontractors should be easy. You might think that there would be a long line of subs standing at your office door in hopes of catching a scrap of work. This could be the case, but it probably won't be. Even if you do have a few eager beavers on your doorstep, you have to be careful. If they are really good subcontractors, they should not have much need to come to you in search of work. Experience has shown me that the best subcontractors are working, not looking for work. This is why cruising job sites can be an effective means for finding subcontractors to bid your jobs. Ride through projects and jot down the names and numbers found on the trucks at the jobs. The fact that the subs are on a job gives you two advantages. One, they are working, and this is a good sign. Two, you can inspect their work to see if you want them to work for you. The downside is that they may be too busy to take on your work.

The advertisements in your local phone directory is another way to solicit bids. You can thumb through the ads and call contractors at random. The companies listed in the phone book will have been in business for at least a little while, or they would not have had time to get their ad in a book that comes out only once a year. Of course, they may have hit the timing right and have just one month of experience in business. Wherever you find subcontractors, you have to check them out to see that they are licensed, insured, possibly bonded, and that they do good, dependable work. Remember, your name as a general contractor is riding on the quality of the work that your subcontractors do.

Bid packages

Once you have a pool of subcontractors to pull from, you should offer bid packages. A bid package is simply a set of information that is consistent with the information given to all other subcontractors. When you give subcontractors a formal bid package to work with, you can be sure of comparing apples to apples. It is far too easy to have discrepancies in bids when detailed bidding instructions are not given.

What should you include in a bid package? It depends on the type of job that you are putting out to bids. If you have a floor plan or detailed blueprints to work with they should be included in the package. A detailed list of specifications belongs in your bid packages. Basically, any information that may affect a subcontractor's price should be included as part of the bid package, and all bid packages should be the same for each trade that they are being distributed to.

When bids come in

When bids come in, be prepared for some unexpected results. If you gave out identical bid packages, you should receive competitive bids. This is not always the case. Even on small jobs I've seen bid prices vary to a point of being outrageous. It is not uncommon for general contractors to throw out the high bid and the low bid before they even start to assess the prices. Standard procedure is to get at least three bids on a job. I recommend that you get at least five bids if you are unfamiliar with the contractors who will be giving the bids. Once you build a dependable stable of subcontractors, you can cut back on the number of bids you get. However, don't fall into a rut with one subcontractor and allow prices to creep up on you without realizing it.

Once you open bid responses, you will have to sift through the bids for reality. Subcontractors are supposed to follow the instructions in your bid packages, but they often don't. For example, if you specify a particular brand of plumbing fixture and state that no substitutions may be used in the bidding process, you would justly assume that a plumbing contractor would comply with your request. Don't count on it. Check all aspects of your bids carefully. If a subcontractor does make a substitution on materials, it can have a substantial effect on the overall price. While the price may be lower, it may also be useless if you or your customer is unwilling to accept the substituted material.

Unfortunately, a lot of subcontractors, especially the ones accustomed to doing small-scale residential jobs, don't pay a lot of attention to the rules of a bid package. Contractors who do large jobs are used to living up to the terms of a bid package. Why is there a difference? Many general contractors don't get formal on small jobs. Therefore, subcontractors tend to be lax in reading the fine print, so to speak. It's not that they can't, just that they don't. I don't intend this as a blanket statement on all subcontractors who do small residential work, but it holds true often enough for you to be aware of the potential problem.

When you go over the bids you receive, take notice of the attention to detail in the proposals from subcontractors. You can't judge field work from office work, but you can get an idea of how professional the subcontractors are. For example, if you get a proposal presented on a stock form where the subcontractor's name is either written in or stamped on the proposal with a rubber stamp you will garner one opinion. If the same information was submitted on a custom-printed, personalized proposal, you might feel somewhat differently. It's not fair to make a full opinion based on a proposal, but you can get a good feel for the type of contractors you might be working with after studying the submissions.

Pay attention

Aside from professionalism, there are some key aspects of proposals that you need to pay particular attention to. The first order of business is to see if the subcontractor is insured properly and has provided the name of the insurance company. Personally, I would not work with any independent contractor who was not insured. Some subcontractors will attempt to get you to hold money out of their bid to cover their insurance. A few contractors do this, but I don't recommend it.

Did the contractor bid the job with materials that you specified? If substitutions have been made, you need to be aware of it. Don't hesitate to call the subcontractor and inquire as to the brands and model numbers of fixtures being bid, if such information is not in the subcontractor's proposal.

Is the subcontractor willing to accept your payment terms, as outlined in the bid package? I make it a habit to limit contract deposits to a minimal amount. It is not unusual for subcontractors who have not worked with me to request one-third of the contract amount in the form of a deposit at the time of signing. This is not acceptable to me. There have even been contractors who have asked me for half of the contract amount up front. This is ridiculous, and it happens to be illegal in my home state. You can do what you want with deposits, but I suggest that if you give them at all, and I don't, keep the amount as small as possible. One of the best levers you have as a general contractor is money.

Has the subcontractor agreed to deliver labor and materials, as applicable, within the time frame that you outlined in your bid package? Will the subcontractor accept a clause in the subcontract agreement that charges the subcontractor a set sum of money for every day that the subcontractor is responsible for holding up your work? Go over the proposals closely. Subcontractors are often vague in their proposals. You should want your contracts to be as explicit as possible, and the more that you can deal with at bid stage the less you will have to fight over at the contract stage.

Calling subcontractors in

Calling subcontractors in for a face-to-face meeting has always proved helpful for me. Yes, it takes some time out of the work day, but the results are usually worth it. When you meet subcontractors in person, you can tell a lot more about them. You get to see the type of vehicle they drive, you can assess their knowledge in little word tests, and you can see if your personality will fit well with theirs. I sometimes meet subcontractors on job sites when I'm building a new house, but I never meet unknown contractors at the home of a customer. By making a habit of meeting my subs at the office first, I can screen out the ones that I feel may tarnish my image with my customers. This can be important.

Once you have brought subcontractors in for a meeting, you need to ask questions. Some of the questions are pretty standard. For example, you might want to know how long the contractor has been in business. Does the contractor work alone or are employees or subcontractors used? Can the subcontractor be reached by pager or cell phone? Are references available for the subcontractor? When I get references I prefer jobs that are in progress. I like to see work being done and talk to contractors or homeowners who are presently using the subcontractor who I'm thinking of working with. This gives me much more confidence than calling the names of three people who could be relatives of the person. You should check references every time you bring in a new subcontractor. I will admit, however, that I don't always do this. If I get a good feeling about the person, I might try the subcontractor without references. It will not take long to find out on the job if the subcontractor is any good, and my contracts allow me to eject a subcontractor and bring in a replacement if the work or performance of the contract is not to my satisfaction.

Cutting a deal

Cutting a deal that you are happy with may be harder than you think. Cheap subcontractors are not always good ones. The best subcontractors stay busy most of the time, so they have no need to accept a bid price that is lower than it should be. Can you drive a hard bargain for cut-rate prices? You probably can, but you must be careful not to do more harm than good when doing so. What do I mean by this? Saving a few bucks on a job will not prove profitable if the job is late in completion, turns out poorly, or if your customers are unhappy with the performance of your inexpensive subcontractor. There are clearly times when you would be much better off to pay a premium price for top-quality service.

The next chapter is going to address the options for getting rock-bottom prices with suppliers. I've found that it is much easier to drive a bargain that turns out to be a true bargain with suppliers than it is with subcontractors. Why is this? Suppliers will be offering you the lowest price on a known object, such as a light fixture. When you start beating subcontractors down on their prices, you may get inferior work. There is no effective way to compare subcontractors equally. True, you can see that they are all bidding the same work, but you have no way of knowing how they will go about doing the work and what quality of workmanship they will give you. This is what makes a bid-for-bid comparison for labor difficult to judge.

There are a few ways to get even the best subcontractors to consider working for you at lower prices. One of the best ways is to commit to the subcontractors to give them all of your work within their field until there is a problem with price, service, or quality. It is comforting for a subcontractor to know that a volume of work will come from you, and this is a justifiable reason to work for less money. When time is not a factor, you can offer jobs to subcontractors on a fill-in work basis. This means simply that the subcontractor will work other jobs until there is a rainy day, a material delivery that goes sour, or some other event that leaves the sub with no other place to work. There are times when this type of arrangement works well, but bathroom remodeling usually doesn't allow for so much freedom in the time allotted for completion.

Prompt payment is something that all contractors enjoy. It is not unusual for subcontractors to find themselves waiting a long time for payment. General contractors get into cash-flow crunches and have trouble making ends meet sometimes. If you can guarantee to pay your subs at the time of completion and acceptance, and mean it and stick to it, you might get a little break on the contract price.

There are not many good ways to drive prices down on desirable subcontractors. The fact that they are good at what they do is why their services are in demand. If you want top-quality work, you will probably have to pay well to get it. This can still be a bargain. A good job at a fair price is much better than a bad job at any price.

Dealing with subcontractors

Dealing with subcontractors should go smoothly if you follow some basic guidelines, but there are sure to be times when you will wish you were in some other line of work. There have been times when my business has used in excess of 100 subcontractors and vendors. Whenever you deal in large numbers, you have to be well organized if you hope to profit. The same principles apply to working on a small scale. Paperwork is one of your best means of staying on top of subcontractors. I know, you hate the word "paperwork" and your subcontractors simply won't tolerate it. Bear with me on this one, because I've been there and done that.

As distasteful as paperwork may be, it is an effective tool when used properly. Subcontractors have to be managed. Allowed to work at will, there is no telling what the results might be. If you are a general contractor, you have to call the shots and take responsibility for them. This is as much a part of your job as making sales, building walls, or paying bills. If you take the time right now to create a system for working with your subcontractors, you will save time down the road and you should reduce your headaches considerably.

Few contractors like to take time out of their production schedule to sit around and fiddle with forms. I know the feeling, but believe me, it does pay off. You will need some help from an attorney for some of the forms, but you can do many of them yourself. There are books available that are written about nothing but forms. We don't have that much space here, but let me give you a quick rundown of some of the most important forms that you can use to manage your subcontractors.

Bid request

One of the first forms needed is a bid request. This is a simple form that you can make yourself. It asks subcontractors to respond to your solicitation for bids. The form can be a letter; it doesn't have to be fancy. Use a bid request any time you want a subcontractor to give you a price. Keep a copy of all forms mailed out. This will tell you which subcontractors responded and how quickly they reacted to your request. After a few mailings, you will be able to determine which subcontractors should be taken off your bid list.

Contract

Where would a general contractor be without a contract? Of course you need contracts for your subcontractors. Some subs will come to you with a contract of their own, but you are in a better position if you supply the contract. Ask your attorney to help you with this form. It is illegal in most, if not all, states for anyone other than a lawyer to create a contract, since it is a legal document and the preparer is practicing law.

How detailed your contract becomes is up to you and your attorney. My contracts are very detailed with every conceivable situation addressed. Some subcontractors don't like my contracts, because they give me a lot of control and power. But, I've never had a sub refuse to sign one of my offers. Your lawyer can help you a great deal in figuring out what you want in your contract.

Change orders

Change orders should be created by your attorney or you can buy them from catalogs or stores that sell contracting forms. Whenever you make a change from your original contract you should use a change order to document the alteration. This can save a lot of confusion, anger, and money in legal battles as your career progresses.

Lien waivers

There are two types of lien waivers—short-form waivers and long-form waivers. Most lenders require lien waivers to be signed and on file before disbursements of construction funds will be given. As a protection to you and your customers, you should have your subcontractors sign lien waivers whenever you pay them. Again, you should consult your attorney on the creation of a lien waiver.

Other forms

There is a host of other forms that can make your life easier when dealing with subcontractors. Unfortunately, we don't have time to delve into all of them in this book. I wish we did, but we don't. I've got forms for every imaginable purpose from paint colors to satisfactory work acceptance to code violation notices. Paperwork may seem like a drag, but it really can make your work much easier once you have a system and know how to use it properly.

Loyalty

Loyalty is a trait that there is too little of anymore. There was a time when employees were loyal to their employers. This was before downsizing, RIFS, and cuts just before retirement benefits could kick in. Employees don't have as much confidence in their employers any longer, and the feelings are not without reason. Some of the same can be said for subcontractors. There are general contractors who put every job out to bids and who will accept only the lowest bids. I don't do this, and I have great subcontractors. Once I have a stable of exceptional subcontractors I pay them well and keep them on my jobs. It costs me a little more on a per-job basis, but I believe it saves me money and frustration in the long run. If you can build a rapport with your subcontractors and maintain a fair business approach to your jobs, I feel that you will prosper from your actions.

Negotiating with suppliers

NEGOTIATING with suppliers is one way that a general contractor can save a lot of money. Saving money on supplies translates into making bigger profits on a job. Another way to look at cutting material cost is that it allows you to bid a job at a lower price and make the same amount of profit. Either way you look at it, you will do better if you learn how to keep suppliers in line on the prices that they give you. There's no real magic to making suppliers give you better prices, but you do have to know a few techniques to make the dream come true.

Bathroom remodeling doesn't always require a lot of material. It's not like building a house. The fact that you will not be buying in bulk hurts your chances for demanding lower prices, but don't give up. If you do enough remodeling work, your end-of-the-year totals with suppliers may be more than enough for them to want your business badly. The key to getting good prices is often volume buying power, but this is not the only way to reduce your costs and increase your profits.

What is your standard procedure for getting supplier prices? Do you give the supplier a set of plans and ask for a take-off and a price, or do you supply a materials list to the supplier? If you prepare the list of materials yourself, you will get a better perspective on what the supplier is charging you. It's possible that the supplier might give you a little break on your cost since you did the take-off, but this is something of a stretch. Suppliers like to work with contractors who do a lot of business. The discount pricing plans are often based on a contractor's volume. This can be true of lumber, lights, or plumbing fixtures. How much you can get a supplier to come down on prices depends on many factors, so let's look at some of them.

Your volume

Your volume of material purchases may not be a lot in the eyes of a major supplier. The best street prices often come from the larger suppliers. Why do you think this is? Right, they are big suppliers who buy in huge bulk amounts, so their suppliers give them better pricing that they can pass on to the customer. This is just the way it is. But, there are exceptions that you can find, and we will get to them soon. For now, let's concentrate on your volume of purchases.

Let's assume that most of your business is kitchen and bathroom remodeling. How many sheets of drywall does it take to do a bathroom? Not many, especially when you compare it to the number of sheets it takes to do a complete house. Now you walk in and want a price on a few sheets of drywall while there is a contractor at the same counter asking for a price on two truck loads of drywall, who do you think will get the better price? My guess is that the volume buyer's price will be much lower than yours. This is not always the case, but it usually is. How can you compete with such a big buyer? You can't, at least not directly. Okay, so you're between a rock and a hard place, right? Not yet; you still have options.

It is impossible for the little contractor to compete with the large contractor in sales volume. With this being the case, you need another angle. One approach might be to work with a smaller supplier. Big corporations don't usually deal with small suppliers because the suppliers can't provide them with the quantity of material needed. This puts the little supplier in a similar situation to the little contractor. When the two little businesses work together they have a chance of a win-win situation.

You can probably convince a small supplier to cut your prices if you buy all of your material from the same supplier. There is a problem with this approach. While your discount amount may be greater the price of the materials is probably higher to start with, so you are in an equally poor position. So, what can you do?

As a kitchen and bath remodeler you don't use a lot of drywall, but you do buy a lot of cabinets. Ah, now there's an idea. The other contractor buys mostly drywall, but you buy light fixtures, windows, doors, skylights, cabinets, plumbing fixtures, flooring, lumber, and so forth. Your per-job quantities are not tremendous, but you do a lot of jobs so your business will add up. If you talk to a few of your contractor buddies, you might even be able to make the deal sweeter. I'll tell you more about this soon.

The straight approach for a lower price is to shop many suppliers until you find one who is willing to deal with you. This could take some time, so don't get discouraged. I mentioned earlier about letting the supplier do the take-off as compared to you doing your own take-off. Do your own and provide a detailed list of what you want prices on. This is a little more work for you, but it's worth it, and I'll tell you why.

When I was first getting started as a home builder I let suppliers do my take-offs for me. It saved me time and I liked the fact that the suppliers were catering to me. However, it didn't take me long to figure out that I was losing money by trying to save a little time and effort. Let me give you a quick example of how this happened.

I remember a group of three spec houses I was preparing to build. When I wanted prices on materials, I distributed blueprints to four suppliers and asked for prices. The spread between the companies was enormous. There were many reasons for this. One company didn't bid some of the material because they didn't carry it. Another company put in a low allowance for all the cabinets and did not do the take-off for the lumber properly. The other two companies were close to right, but they were both off the mark in both the materials needed and in the overall pricing range. For example, company A was giving me a great price on lumber and drywall, but their prices on shingles, siding, and cabinets stunk. Company B would give me a good price on this or that, but their prices on other materials were out of line. I learned two things very quickly from this experience.

If you want your supplier bids to be as equal and as competitive as possible, you have to provide a list for them to bid and the list must contain detailed specifications on brands, grades, and model numbers. The other thing I learned was that it would be difficult to deal with one supplier for all of my needs. The reason for this was that some suppliers got better prices on some materials than others did. For example, I might buy all of my siding from one company, all of my lumber from another company, and so forth. This is more of a hassle than dealing with a single supplier, but the amount of money on an entire house can reach into thousands of dollars. I prefer to do all of my dealings with just one or two suppliers, so I set out to find a way to make myself happy without going poor in the process. Guess what? I did find a way, and I will share it with you.

I compiled all of the numbers from all of the suppliers and grouped them by categories. Then I went to the supplier that I liked the most and sat down with the general manager. The manager and I went over my situation. I explained how I wanted to deal with as few a number of suppliers as possible and that this particular supplier was my first choice. Then I asked the manager to meet the prices of the other suppliers in return for all of my business. He couldn't come close on the siding price, but he got in the ballpark on everything else. Then I got him to guarantee the prices for the length of time that it would take my crews to build the home. He agreed and I agreed to work only with him, except for the siding.

I was able to do the deal I just told you about because of my volume business and my approach. Could I have done it with a lower volume? Frankly, I doubt it. So, you're still wondering how you can do it when what you do is remodeling and not building. Read on my friend; I will show you.

Tiny suppliers

Tiny suppliers exist even in large cities. If you live in a rural area, there are probably many tiny suppliers you can work with. These suppliers usually have to pay more for what they buy and sell, but some of them manufacture their own products. If you find a supplier that creates the product being sold, you are in a good position to deal for a low price and high quality. Let's take a cabinetmaker as an example. Since you do so many kitchens and bathrooms, both of which require cabinets, you can offer a small-scale cabinetmaker a lot of volume. Here's your first answer. Under the right conditions, you may be the cabinetmaker's largest customer. This puts you in a power position. Not only can you get your products for a good price, you can probably get some special treatment along the way. Cabinets were easy, since they are what you need the most of, but what else is possible?

Let's say that you normally install vinyl flooring in your remodeling jobs. This is standard procedure. The problem is, you don't use enough vinyl to capture the attention of a flooring supplier. Eliminate the supplier and deal through your installer. An installer does a lot of floors and buys a lot of flooring. Some installers, like the one I use, have their own distributors, so they are basically buying factory direct. This gives the installer a much higher profit margin on most jobs. If you make it a contingency of getting your installations to sell you the flooring at say 10 percent over cost, you can save a lot of money. The flooring contractor might balk at this, but there will be some installer who will jump on it. Paying 10 percent over the installer's cost could save you thousands of dollars over the course of a few jobs. It's well worth looking into.

Drywall is a tough one, but you don't use much of it and it doesn't cost a lot, so it's really no big deal. You might be able to lump your drywall, plywood, and lumber deals all with one supplier and get better pricing. If you install a lot of windows, doors, and skylights, your chances improve, since most suppliers who sell any of these materials, sell all of them. Light fixtures and plumbing fixtures might also be sold by the same suppliers, but I suspect you can get a better price out of your plumber and electrician than you can out of the building supplies store.

Since remodelers frequently deal in small quantities of many types of materials they can often get better prices out of their subcontractors than they can out of suppliers. This chapter is about suppliers, but you should consider buying through your subcontractors for much of what you use. Convince your subs that in order for them to get the labor portions of your job, they will have to sell you the materials they use at the lowest rates possible. Some will and some won't, but it's worth a shot.

Co-op purchases

Co-op purchases are another way to put a dent in high material prices. If you have some buddies in the contracting business you can suggest that you all bind together and deal with a single supplier. Assuming that you have a decent relationship with your competitors this is a very real possibility. One remodeler may not buy enough material to attract the attention of a supplier, but when there are three, four, or five accounts on the line, a supplier might become much friendlier in terms of prices. This type of deal will probably require a commitment from all contractors to deal almost, if not entirely, exclusively with the same supplier, but the prices may well warrant such action. The downside of this approach is that it takes a combined effort from you and your competitors, but everyone can come out a winner if the deal is cut right.

Out-of-town suppliers

Working with out-of-town suppliers is one way to drive hard bargains on material prices. While not always practical, dealing with suppliers who are outside of your normal work area can produce good results. I've had suppliers from other states work with me and deliver right to my jobs. If you work in a progressive area where prices are held up due to supply and demand, consider dealing with suppliers in more distant locations who don't have the same demand for goods that local suppliers do. Some out-of-town and even out-of-state suppliers ship once a week to progressive areas. I have a regular supplier in another state who sends a truck my way twice a week.

If you find suppliers in out-of-the-way locations, they may be happy to offer you low prices in return for steady business. Delivery can be a problem, but I've known a number of suppliers who would ship or deliver for prices that were lower than my best local prices. For example, I get my plumbing materials shipped from a home office in Jacksonville, Florida to my shop in Maine. As long as my order is over $500, I don't pay any shipping charges. The prices I get from this supply range from 10–33 percent less than what local suppliers will sell to me for.

Using out-of-town suppliers can work whether you live in a small town or a large city. Until you solicit bids you won't know who will sell for less. It might be that big-city suppliers will ship to you for a lower price than your local wholesaler. You might find the reverse to be true. Until you ask for pricing from all available sources, you will not know where your best deal might come from.

Become a distributor

Think about becoming a distributor for items you install a lot, such as cabinets. Hooking up with a factory-direct source can save you a lot of money if you have enough volume to justify doing such a deal. Cabinets are probably the only area that kitchen and bathroom remodelers would fit into, but it's worth looking into. I've never gone this route, but I know subcontractors who have, and they save a lot of money on their materials. If this idea interests you, contact manufacturers directly and inquire about a distributorship. Don't count on getting low prices just because you are awarded a distributorship. Like regular suppliers, manufacturers often rate their distributor discounts on a volume basis. You could wind up paying more for a cabinet as a distributor than what you could buy it for from a high-volume wholesaler.

Ask for a bigger discount

Sometimes the easiest way to get a bigger discount is to ask for it. I've done this many times over the years and it has worked often enough to keep me doing it. My first experience with this approach was almost an accident. I was on good terms with a lumber supplier and had been getting fair prices. One day I was joking with the counter help and said I wanted to talk to the manager for a lower price. I knew the manager, but I had no idea my sarcastic remark would be taken seriously. The manager came out and called me into his office. He offered me a cup of coffee, which I declined, and then we went to work on negotiations. By this time, I was semiserious. I simply asked the manager for an additional 2 percent off my present discount due to my good payment history and my loyalty to the supplier. My request was granted without argument. From that day on I started saving an extra 2 percent, just because I asked for it. If you don't ask for what you want, you probably won't get it.

Buy in volume

If you can afford to buy in volume and sit on what you have in inventory, you can save some money. However, you don't want to sit on the supplies too long with your money tied up. The trick here is to buy materials that you know you will use within say two months. If you have enough volume to buy faucets by the case, cabinets in a lot, and so forth, you can expect lower prices. You will, however, need warehouse space and your money will be tied up in inventory. This is not a good move, unless you have an established demand that you can count on. If you buy a lot of materials and the demand drops off, you're stuck with boxes of stuff that you can't sell well.

Searching

The most effective way of getting ideal pricing is to keep searching and negotiating. Phone books are full of suppliers. New suppliers pop up from time to time, and they are hungry for business. Established suppliers fall into slumps occasionally and are willing to make better deals for new customers. The search for low prices is an ongoing process. However, if you find one or two suppliers that you are happy with and on whom you can depend, their qualities may justify slightly higher prices. You have to make the decision of whether to go for the lowest price or to stick with a known supplier. Personally, I tend to stick with the suppliers who take the best care of me in terms of delivery, honesty, and so forth. A low price is always welcome, but service and dependability is worth a lot in the contracting business.

Quick decisions

WOULD you believe that a quick decision on a job can cost you money, your reputation, and a loss of business? Sounds extreme, but it's true. If you act, or react, too quickly to a situation on a job, you can find yourself in a world of trouble. Customers may get angry. You could lose valuable subcontractors. A wrong decision could cost you hundreds, if not thousands, of dollars. The risk on making on-the-spot decisions can be great. You may be thinking that you're immune to such a problem, but I doubt that you are. Almost anyone can be prodded into making an on-site decision that will be regretted for years to come. It could be a customer asking you for a price to add a new window while you are remodeling a bathroom. You could find your problem to be with an electrician who wants an instant answer on where to rough-in a new light. The sources of trouble can be numerous, and a busy contractor can be susceptible to making a mental mistake.

How often do contractors get in trouble by talking without thinking enough? Pretty often. It happens with subcontractors and general contractors. The mistakes are usually avoidable, but not always avoided. A prime reason for the slips of the tongue is preoccupation with other matters. I've made the mistake more times than I'd like to admit to, and I've seen it happen to others over and over again. For some reason, it's almost as if busy contractors just can't help but put their feet in their mouths sometimes. When I first suffered the consequences of such mistakes I shrugged them off as a cost of doing business. Then when the situations happened more and more often, I took a harder look at what was going on and decided to put a stop to it. Maybe you have never stepped into a verbal trap. If so, you are either lucky or darned good. But most contractors have made or will make mistakes with their poor decision to talk off the top of their heads, and this chapter is intended to point out why you shouldn't and how you can avoid it.

My first recollection

My first recollection of a situation where talking off the top of my head could get me in trouble came when I was on a job estimate as a subcontractor with a general contractor. The job was a simple, little remodeling deal involving a laundry room. I was young and somewhat inexperienced, but even I could see that the job didn't amount to much. The man I was with was in charge, so I just observed as he and the homeowner talked. After Roy, the general contractor, and I had finished our inspection of the job, we found the homeowner to say that she would hear from us soon. The woman wanted a price right on the spot. At the very least, she wanted a ballpark price. If I had been in control, I'd have probably gone to the truck, figured the deal, and given her a price. Roy explained to the woman that he couldn't give her a price until he got back to his office and calculated the job. This seemed silly to me. The woman wanted a price so that she could make an on-the- spot decision. It seemed strange to hang a hot customer out on a limb for awhile. Nevertheless, Roy wouldn't budge on his decision to withhold pricing information until he had a chance to run the number back at the office.

I was riding with Roy on this particular estimate, so I picked his brain a little on the ride back. When I asked why he wouldn't give the lady a price on such a simple job, he had various explanations. The first reason he gave was that it wouldn't look good to spit a number out too quickly. Roy said that customers feel better if they think you labor over a price for hours. The reason for this was said to be that a quick price is a high price and a calculated price is a fair price. In a way, this made sense. But I still didn't buy it completely. I asked Roy if he already had a price in mind. He did, and he shared it with me. My estimate of the job was very close to his, and neither of us had sat down to figure it. If two independent contractors came up with bids so close together so quickly, what would be the harm in sharing the price with an anxious homeowner? I still didn't understand the stall tactic. Roy went on to explain that there might be something about the job that we were both missing. He said he wanted time to sit down and chew on the bid a bit before committing to a price. I understood what he meant, but I still didn't see the value to putting the lady off on her price when it was such a cut-and-dried deal.

It didn't take long for us to get back to Roy's home office. He invited me in and I gave him a price for my part of the work. Roy nodded and said it sounded okay, but that he'd have to get back to me on it. I wasn't any happier than the homeowner. My price was good, and I wanted an acceptance on the spot, but Roy was stubborn and wouldn't give me one. We had worked together before, but not a lot. I was starting to think that Roy was just a difficult contractor to deal with. As it turned out, he got the job and so did I. The job sold for about what we had thought on the ride back to Roy's house the night of the estimate visit. So if everything worked the way I thought it would, why did it take so long to put it all together? Roy was cautious. He was older and much more experienced than I was at the time. His position and actions were justified, as I learned later in my contracting career. Youth and inexperience blinded me at the time, but I came to realize how smart Roy was in not giving a price or accepting my bid right on the spot.

It's been about twenty years since I took that ride with Roy, and his words of wisdom still ring in my head sometimes. Being young and impatient back then, I didn't appreciate the slow, businesslike approach that Roy and some other general contractors took. It was years before I realized how right they all were and how wrong I'd been. Now the tables have turned. I'm the older contractor with young ride-alongs who don't understand my procedures. It has taken years for me to gain the insight to the value of listening rather than talking and keeping my mouth shut until I'm sure of what I want to say. Even with this experience, I still mess up sometimes. Maybe it's my personality that gets in the way. I don't know why I still do it on occasion, but I do still talk quicker than I should at times.

Don't talk; listen!

One of the best rules in sales and in customer satisfaction is simple; it is this: don't talk, listen. I came up with this saying in one of my seminars for increasing in-home sales. Many salespeople talk far too much. You may not think of yourself as a sales professional, but you have to sell jobs, and that puts you in the sales category. Customers want you to answer their questions, but they also want a chance to speak their mind. If you go on a sales estimate and do all the talking, you will not find out what the customer wants. You have to be willing to listen in order to understand how you can best serve your new customer.

What does sales have to do with quick decisions? A lot, really. If you know how to use sales skills, you can put them to use in many situations. The same skills used to sell jobs can be used to calm customers you've already sold. Since you are doing the work, you must have made some impression on the customer. Some customers go strictly with the lowest bid they receive, but most people want to work with a contractor they like and trust. If you gained the customer's confidence in the beginning, you should be able to maintain it through disagreements. Of course, you will have to act properly to accomplish this goal.

It doesn't matter whether you are taking notes about what a customer wants done, selling a job, or getting a customer who is upset to calm down, you need to do more listening than talking. This is difficult for some people to manage. It's not uncommon for contractors to talk almost nonstop. Then there are contractors who barely utter a word. The happy medium between these two extremes is where you should be. Offer your customers small talk, but save the heavy stuff for the customers to talk about. Make notes, either mental or written, so that you can assess what it is that the customer is really saying. People often don't speak their mind clearly, and you may have to read between the lines. One of the best defenses you have is to be a good listener.

Let's set the stage for an example and see how you feel about the scene. Assume that you are nearly done with a bathroom remodeling job. The work has gone well up to this point. You are at a stage where fixtures are being put in and the plumber scrapes the new vinyl flooring with a vanity cabinet. It's a stupid mistake that shouldn't have happened, but it did. You get a phone call from your customer around lunchtime, when the customer came home to check the job and discovered the gouge. Your customer, who has been very nice throughout the job, is livid. All you can hear on the other end of the phone is how upset your customer is. What do you do now?

You should listen and keep your mouth shut unless asked specific questions. More often than not, the customer is blowing off steam and working through frustration. If you can wait out the storm, the customer will probably run out of gas and slow down, so that you can talk in a calming way. Should you break into the conversation and tell the customer that you will replace the floor? Absolutely not. Don't stop the customer's ranting. Wait until the customer pauses and then offer to come out to the job to see what damage has been done. Don't agree over the phone to any replacement, repair, or other arrangements. Get to the job and get face to face with the customer. Don't offer any solutions until you have a full understanding of what is going on.

Once the customer slows down on the phone, offer to go out to assess the complaint. Try to put off your trip until the customer goes back to work. It's to your advantage to see the damage in a quiet time, without the customer hammering you the whole time that you are looking at the problem. This may not be possible. The customer may insist that you come out immediately. If you can, go ahead and get to the job. Putting off the customer under these conditions could worsen the situation.

When you get to the job, look at the damage. It may not be nearly as bad as the customer has indicated. Keep in mind that even if the floor is cut badly, you are almost at the end of a job that has gone well and one that will result in a good reference if you handle this problem properly. It may be worth paying for a new floor to keep the customer satisfied, even if a patch job would work nearly as well. Customers do not like patches in new work, and your reputation could suffer greatly from a greedy approach on your part.

Let's assume that your customer insisted on your immediate attention, so you have to meet the customer on the job. Expect some more anger when you arrive and just prepare to take it. Let the customer vent as much as necessary. The more frustration you allow the customer to get rid of now, the easier your deal will be later. Once you are in the house, ask your subcontractors to take a break. It's not wise to fill a room with people when you are dealing with an irate customer. Get your subs out of the job, so that they will not be exposed to the customer's rage. The last thing you need right now is some plumber telling your customer that the ding is no big deal. Clear the room and then look the problem over with the customer. Even if the customer's opinion of the problem is ridiculous, don't let your amusement or distaste show. Look at the floor and let the customer talk some more.

At some point, the customer will slow down and ask you just what you plan to do about the problem. An alternative approach for the customer might be to tell you what you are expected to do. Just listen, don't talk yet. It's okay to nod some, but don't agree to anything. When the customer runs out of threats, complaints, and other comments, you get your turn. Explain to the customer that you will talk to your flooring installer and your plumbing contractor about the problem. Apologize for the incident and assure the customer that you will see that the job is finished to satisfaction. Don't make any firm commitments on exactly how this will be done. Buy yourself some time to think and to talk to your flooring installer. Most customers will accept this type of response.

When you get back to your office, go over the facts and decide on a course of action. Does the floor need to be replaced to make a decent job? Should you charge your plumbing contractor for the cost of replacement? Was the plumber negligent? Can the floor be cleaned or patched to hide the incident perfectly? Work through your options and make a decision. Once you have done this, call your customer and apprise the person of your decision. Anything less than a total replacement may result in a hostile response. So if your flooring sub says the floor can be fixed without any apparent difference from a new one, offer the customer a satisfaction-guaranteed patch. This could be buying trouble for yourself. Regardless of how good the patch is, the customer may not be willing to accept it. Under the conditions of this example, I would be inclined to replace the flooring, either at my expense or at the expense of the plumbing contractor, depending upon the actual circumstances causing the trouble. If I was very fond of my plumbing contractor, I'd probably either pay for the replacement out of pocket or offer to split the cost with the plumbing contractor.

Customer relations

Customer relations are extremely important to your business. If customers don't like you or your company, you will not benefit from the great gains that can come from good word-of-mouth advertising. You can't buy better advertising than satisfied customers. This means that you have to go to extremes to keep your customers happy. Whenever possible and feasible, you should do this. There are times, however, when customers are simply impossible to satisfy. When you run into this type of situation, you have to know when and how to cut your losses and move on. It's always a shame to lose a customer, but some customers are better lost than kept.

I've been in business for a very long time, and I've dealt with countless customers of all types. If you stay in business long enough, you will run into people whom you just can't please, no matter what. When this happens, you need to recognize the situation as soon as possible and get out of it just as quickly. Never get in a direct confrontation with a customer if you can avoid it. Part of your job as a service provider is to take grief from disgruntled customers. If you do your job well, the occasions when this happens will be rare. But when you have to take the guff from a customer, try to do it with an understanding look on your face. Nobody would expect you to smile while you are being beaten up verbally, but give the customers a chance to get their anger out of their systems. There will be many times when this is all that is required to settle a problem. On the few occasions when you can't reach a satisfactory conclusion, do what you have to do under your contract and walk away. You will serve no good purpose by engaging your angry customer in an ongoing battle.

Taking control of a situation

Taking control of a situation is part of your responsibility as a general contractor. This is a fact whether you are dealing with customers of subcontractors. The customer is your boss, but you are in charge of the job, and that should put you in control, at least to some extent. Most of your control, if you want to call it that, will be exercised over your subcontractors. It's acceptable for you to basically tell subcontractors what to do, since you are the one paying them. By this same token, customers expect to tell you what to do, since they are paying you. They may not tell you how to get the job done, but they do have a right to get what they contracted for. Savvy contractors know how to control customers without the customers knowing that they are being controlled. This skill is developed over time, and it's not easy to teach in a few paragraphs. But I will point you in the right direction.

▲

113

I'm not an expert on human behavior, but I have logged a lot of hours dealing with people on remodeling jobs. The best advice is what I've already given you: you should let people talk their feelings out before you even attempt to deal with them. This requires patience, and it can be difficult to do. However, you must take control of your own emotions before you can hope to control the emotions of others. Ah, here is the second most important piece of the puzzle. I just said it, but I'll say it again, you have to be in control of yourself if you hope to control any situation. Let me expand on this a bit.

There was a time in my life when I responded to emergency calls in an ambulance. When you arrive on an accident scene and are expected to help people you cannot panic, no matter how bad the scene is. Not all people have this type of self-control. I learned at an early age how important maintaining my composure was if I was to get my
job done. The same logic carried through into my construction and remodeling career. When I come into contact with an irrational customer, I just settle in and wait for the storm to pass. Doing this is much easier now than it was in the early years of my career. Youth has a way of pushing patience out of the way. Maturity is a wonderful thing; it's just too bad that you can't have experience and maturity at an earlier age so that you could make more out of it.

Let's assume that you can train yourself to be patient, an attentive listener, and a person who doesn't fly off the handle when someone is criticizing you. If you reach this point, you are well on your way to becoming a crisis manager. General contractors do have to have crisis-management skills if they want to maintain a long list of happy customers. Very few jobs go off without any problems. In my experience, most of the problems with remodeling work have come during about the last 15 percent of the job. I don't know whether people lose their patience, whether subs are anxious to finish up and get out of the job, or if the stuff just hits the fan near the end for some other reason. I do know that on my jobs I have more problems at the end than I do in the beginning and in the middle. This is good, in a way. If the problems hold off until near the end, they are easier to endure. A job where a customer gets nasty right off would be a long job, indeed.

I'd like to be able to tell you that I'm perfect and that my jobs never sour. While I could tell you this, I would be lying. Every contractor faces a certain percentage of problems on jobs. The more jobs you do, the higher your odds are of coming into contact with difficult customers. And it's not always the customers who are the problem. Subcontractors can be a big source of trouble. The same key elements used to defuse a customer should work on a subcontractor, plus you have additional power over the sub in the form of money and future contracts. To round this section out, let me give you a bulleted list of suggestions for handling difficult situations.

➤ Return phone calls promptly.

➤ Don't attempt to hide from a problem.

➤ Buy yourself some time if necessary, but face your responsibilities quickly.

➤ Don't panic.

➤ Listen carefully before you talk.

➤ Don't argue; offer alternative options.

➤ Retreat to a quiet place and weigh your options.

➤ Avoid making on-the-spot decisions whenever possible.

➤ Speak in a calm, soothing tone.

➤ Don't allow aggressive body language to betray you.

➤ Avoid laying blame on others; this looks like a cop-out.

➤ Ask questions rather than delivering ultimatums.

➤ Involve the other party in the decision-making process.

➤ Suggest a cooling-off period if you are at an impasse.

➤ Don't make threats.

➤ Avoid saying anything that you cannot follow through with.

➤ Look for sensible solutions.

➤ Try to leave your personal feelings out of the equation for a solution.

➤ Keep negotiations on a businesslike level.

➤ Don't drop to the level of your adversary.

➤ Know how to read a no-win situation and get out fast if necessary.

What is a hopeless situation?

What is a hopeless situation? Fortunately, I've known very few such situations. However, I have been involved with enough of them to give you some examples. My definition of a hopeless or no-win situation is one where the more you do, the worse it gets. You can give only so much before it is no longer worth the effort. Subcontractors and homeowners can both pit you against no-win situations. To remain at the top of your profession, you have to be able to read the signalsof such a situation and react to it properly. Let's talk about the ear-markings of bad situations that you should walk away from.

If you and your customer begin having disagreements before and soon after a job is started, you may be in for long-range trouble. It is certainly a bad sign to get off on the wrong foot. However, some of these situations are a matter of poor communication, and they can be solved with one or two heart-to-heart talks. I wouldn't give up right off the bat, but I would have my eyes and ears open to what the customer might be maneuvering for.

What should you do when a deal gets messy around the middle of a job? Try to work through it. It's best to solve problems whenever you can. If you just can't come to terms that are reasonable, you may have to walk away, but do whatever you reasonably can to avoid this. Allow me to give you a couple of examples of walk-away deals from my past.

I had a married couple contract me to build a custom home for them. This was about 13 years ago, but it's still pretty fresh in my mind. The couple went for financing and were approved. I got a construction loan, in my corporate name, to build the house. Before drywall was ever hung, the wife wanted out of the deal. She decided she didn't want to stay in the area. Her husband still wanted the house, but the wife wanted to go back to her home state. I stalled construction on the home to work matters out. Why did I stop production? Well, if I couldn't salvage the deal I would have to treat the house as a spec house, and I wanted potential buyers to have as much freedom of choice as possible in how the house would be finished.

After many nights of negotiations and discussions I was getting nowhere with the couple. The man wanted the house and the woman didn't. It didn't look like a situation that would resolve itself easily. I was right. Within a week, the woman went out and bought a new car. She also quit her job, and then she called her lender and told them that she had a new debt and was unemployed. The bank rescinded its letter of commitment for money and killed my deal. I had a spec house on my hands and the husband was facing a divorce. What a nasty deal that was for all of us. This kind of stuff does happen, though.

On a subject closer to what we are talking about, I was called in to remodel a bathroom for a man who had lost a limb due to illness. With the individual's new disability, the bathroom had to be remodeled to accommodate his needs. The contract signing went fine and the work was progressing quickly. Along the way, the homeowner found out that his insurance company would not pay for the remodeling of his bathroom. He had been counting on the money coming from the insurance company. I didn't know this when I took the job. My crews were nearly done with the job when the man became super picky. It took some careful probing for me to find the true cause of his dissatisfaction. Basically, I learned that he didn't have the money to pay me so he was trying to delay the job by complaining. I didn't like the news at all, but I did finish the job. Some contractors probably would have walked away, since the guy had not paid any of the draw invoices to date. I felt sorry for him and completed my work in hopes that he would eventually pay me. He never did. I guess I should have walked away, but in my heart I still feel good that I completed the work. Sometimes you just have to go with your gut feelings.

I'll tell you one more story, and then we will move on. There was a couple who contacted me to install a washer hook-up and dryer connection for them. The husband wanted the installation to go in the basement of the house. His wife wanted the connections installed in an upstairs closet. Seeing a couple who can't agree should have been all the warning I needed, but I stayed in the game. The woman won out and we installed the connections upstairs. When we were done, the husband refused to pay me, saying that he had told me to put the hook-ups in the basement. My contract was clear, but he refused to pay. There was not enough money at stake to warrant a trip to court, so I lost out. The man wasn't mad at me or with the work my crews did; he was upset that his wife overruled him. I was the one who paid the price for their disagreement. Since that time, if all the buying power is not in agreement, I walk away.

Some people seem to enjoy confrontations. If you offer viable solutions to a customer and find yourself getting nowhere, be prepared to walk. Setting a stage for jumping every time a customer snaps is not good. Be fair and reasonable. Go to logical lengths to maintain harmony in your jobs. When you run into a customer who doesn't want a resolution, get out. There are times, such as when you are nearly done with a job, when it is better to take the heat and finish the job. You have to use your own judgment on a case-by-case basis.

Ripping out an existing bathroom

RIPPING out an existing bathroom can be a big job. In some cases the work is as simple as removing a few plumbing fixtures. Other jobs entail much more work. Depending upon existing conditions and the degree of demolition needed, the job can consume days of your time. Not only can the demolition process be time-consuming, it can be dangerous for both you and your customer's house. If you don't follow proper procedures, you may electrocute yourself, flood a home with water, or damage other living areas of the home with all the banging that can go on.

There are some safety hazards involved with demolition work. Eye protection should be worn at all times during the rip-out phase, and ear protection may be needed for some parts of the job. The proper clothing and footwear can reduce cuts, scratches, and punctures. There is a good chance some of your demolition work will be done while standing on a ladder, so caution must be exercised to avoid falling.

This book is not intended to teach you on-the-job safety procedures. There simply isn't room to discuss proper lifting procedures and other safety-related topics. But it is important that you know there are dangers involved with doing rip-outs. The dangers extend throughout the job. If you are not aware of how to work safely with tools, ladders, and general remodeling work, consult books on the subject of safety. While I am not going to attempt to teach you safety procedures, I will show you some tricks of the trade. My many years of experience have allowed me to develop some techniques to make remodeling work easier, and I'm going to share these secrets with you. If you are ready to get down to some serious work, roll up your sleeves, and we will get started.

Preparing for demolition

By preparing for demolition work in advance you can avoid a number of problems. One of the first problems inexperienced people run into with demolition work is the mess it makes in the rest of the home. There is a lot of dust and debris involved with demolition, and keeping the mess contained in the room being ripped out is the first order of business.

Disposing of debris

Before you can begin the containment process you must have a plan for disposing of the debris you will create. If the room being remodeled has a window or door that opens to the outside, you may be able to place a trash container near the opening and toss the debris out as you go along. This not only controls clutter in the workplace, it also makes the job go faster. If you have to pile the rubbish in the room and then haul it out to a trash container, you are handling the materials twice. Try to position a trash container where it is easily accessible, and dispose of your refuse as you create it.

If you are working in a room that is on a second or third floor of the home, you may want to build a chute for your trash removal. Set a trash container below the upstairs window where you will be discarding debris. Use framing lumber and plywood to build a trash chute. The chute should have side rails that prevent debris from falling over the sides as it slides down to the trash container. The chute will resemble a sliding board. Have the chute extend from the trash container to the window, and secure it firmly. With the chute in place, you can dump debris out of the upstairs window and have it land safely in the container.

Dust containment

Dust containment will be your next step in preparing for demolition. If you don't seal the room you are remodeling off from the rest of the house, dust will find its way into carpeting and other parts of the home. The work involved with setting up dust containment is much easier than trying to deal with dust all over a house.

All you need for dust containment is some plastic and some duct tape. Seal all doors and other openings between the workspace and the remainder of the home with sheets of plastic. Cut the sheets larger than the openings you are covering, and allow the plastic to extend several inches past the frame of the opening. Duct tape can be used to affix the plastic to the walls of the room you are working in. Keep the plastic on the side of the opening where you will be working, not on the side where the rest of the home is. Tape the plastic to the walls and floor using long strips of tape. Don't leave any portion of the seams untaped. Don't use duct tape to secure plastic to finished wood floors. If there is any risk that the adhesive on the tape will mar a floor finish or damage the finished flooring in any way, you must use an alternative method for securing the plastic. For example, cut the plastic longer than the doorway being sealed. Wrap the plastic around a piece of lumber, like a wall stud, and staple the plastic to the lumber. The weight of the stud will hold the plastic down and in place. By having the plastic wrapped around the wood, you reduce the risk of damage to the finished flooring.

If you must use a door that opens into other living space for access to the room being remodeled, and you probably will, you may want to use an alternative method for sealing the opening. Since pulling tape loose from the walls and floor every time you want to enter or exit a room can be annoying, you might want to get a bit more creative.

To avoid some hassles with coming and going, cut the sheet of plastic to be used for covering an entrance extra large. Tape the top and one side of the plastic to the walls in the way described above. Attach a 2"- x -4" to the bottom of the plastic and roll the plastic around the piece of wood until the vertical fit of the plastic is tight. The weight of the wood will hold the plastic down and eliminate the need for taping it to the floor.

On the remaining side of the opening, tape the plastic to the wall at the top, middle, and bottom, but don't use long strips of tape: small pieces will do fine. There will be gaps along this edge of the plastic. Left alone, these gaps will allow dust to escape the room. To remedy the dust situation without sealing yourself in tightly, hang a second piece of plastic to overlap the lightly taped edge.

The second sheet of plastic should be taped to the wall at the top and along one edge with long strips of tape. The section of plastic that overlaps the other plastic should be taped at the top, middle, and bottom with the use of minimal tape. When this is done, the room is sealed, but you can come and go easily. All you have to do to open the exit is pull the tape from the bottom and middle section of the overlap and inner plastic. The bottom of the main covering will move easily when you push the wood to one side. This type of arrangement is effective in controlling dust while allowing reasonable ingress and egress to the space.

Ripping out a bathroom

What is involved in ripping out a bathroom? If you are taking the room down to bare studs and subfloor, there is a lot of work involved. You will have to work with plumbing and electrical devices, and you may have to work with part of the heating system. These mechanical systems must be treated with respect. The job will also involve removing wall coverings, ceilings, and floor coverings. There are some things to be careful of. Let's look at how the demolition process might go in an average bathroom.

Assume you have a bathroom prepared for demolition and you are equipped with the proper safety precautions, eye protection, and so on. In this job none of the existing fixtures will be salvaged. Where will you start the rip-out? The logical place to begin is with the plumbing fixtures.

Plumbing

Before you begin tearing out existing plumbing, make sure the water supply to the fixtures is cut off. Don't assume a closed valve has the water stopped; some valves fail with age. After you have closed the appropriate valves, test each fixture to see that the water is, in fact, shut off.

Begin by removing the toilet. A screwdriver and an adjustable wrench are the only tools you should need for this part of the job. Flush the toilet to evacuate most of the water. Remove the nuts on the bolts extending through the base of the toilet. If the nuts are seized and will not turn, they can be cut off with a hacksaw blade. Loosen the nut that secures the water supply to the toilet tank. The toilet can now be lifted off the floor and removed. Some water will spill if you have not used a sponge or a rag to remove all water from the toilet tank.

A complete toilet can be awkward and heavy for inexperienced people to handle. In most cases the tank of the toilet can be separated from the bowl, making removal easier on your back. Toilet tanks that are removable are attached with brass bolts and nuts. By putting a screwdriver in the head of the bolt and turning the nuts, the tank should be easy to remove. Sometimes the bolts are stubborn and a hacksaw blade is needed to cut them. If you get the bright idea to break the toilet into pieces with a hammer, be advised, the broken china can inflict nasty wounds when shattered.

I don't recommend breaking plumbing fixtures with force, unless you have to remove a cast-iron bathtub. Old cast-iron tubs are very heavy. They can be removed in one piece, but many experienced remodelers use sledge hammers to break the tubs into more manageable pieces. If you do this, wear protective clothing and eye protection (Fig. 7-1). Also, be advised that the heavy banging may damage common walls of the bathroom and may cause trouble on ceilings below bathrooms. Nail pops are a common problem when extensive force is used during demolition work.

With the toilet out of the way you are ready to move to the lavatory. Disconnect the trap from the fixture; this can usually be done with a

Figure 7-1

Eye protection should be worn during the demolition process. Model Fran Pagurko—Photo by Images International, a division of Lone Wolf Enterprises, Ltd.

set of wide-jaw pliers. The water supplies must also be disconnected. You should be able to loosen the small compression nuts that hold the supply tubes into the cut-off valves without any trouble. Once the waste and water lines are loose, you can remove the basin.

If you are working with a wall-hung lavatory, it should lift straight up, off of its wall bracket. Some wall-hung lavatories are secured to the wall with lag bolts. If you cannot lift the bowl off the bracket, look for bolts securing it to the wall. The bolts should be visible from under the lavatory as you look at the part of the lavatory that is touching the wall. If none are found, exert some extra pressure to remove the lavatory.

When a bathroom is equipped with a vanity and top, the removal process is different. The vanity top is probably attached to the wall with caulking. Run a knife along the joints where the top meets the wall. Look to see if the top is attached to the vanity or simply sitting on it. Remove any screws holding the top to the cabinet and it should lift off. The screws should be found in the four corners of the vanity. Before trying to remove the base cabinet, check to see if it is screwed to the wall. Screws for this purpose are usually installed through a strip of wood or backing that runs from one side of the vanity to the other, below or behind the vanity bowl.

With the lavatory and toilet out of the way, you have more space to work with in removing the bathtub. Before you can remove the bathtub, you must remove the walls that overlap its edges. Strip the walls surrounding the bathtub to reveal the tub edges. Drywall, tile, plaster, or other wall material will be overlapping a built-in lip on the tub. An exception to this is a free-standing, claw-foot tub or similar free-standing unit.

A hammer works well in removing the walls. Don't cut into the walls blindly with a saw: you might hit live electrical wires. Beat holes in the walls with a hammer and pull the wall covering off. If you have your heart set on using a saw, at least open the walls with a hammer and check for wiring and plumbing before using a saw.

When you have the walls around the tub stripped to the bare studs, you can start the removal process for the tub. Remove the tub faucet first, but be sure the water to the pipes is turned off. Unlike the other fixtures, the faucet for the bathtub will not have small supply tubes; it will be connected directly to $\frac{1}{2}$" tubing or pipe. There may be unions in the pipes to make removal easy, but you will probably have to cut the pipes; this can be done with a hacksaw blade or pipe cutters. Remember to make sure that there is no water pressure to the pipes being cut.

The next step is the removal of the tub waste and overflow. From inside the bathtub, remove the screws holding the trim on overflow. If the drain has a strainer on it, remove the screw securing the strainer and expose the cross-bars in the drain. Using two thick-shaft screwdrivers, insert the screwdrivers into the drain and cross the shafts. By creating an "X" with the screwdrivers you will be able to loosen and remove the drain. Turn the drain counterclockwise to unscrew it.

When all the plumbing connections are loose, you can remove the tub. If the bathtub is a one-piece tub-shower combination, it should be secured to the stud walls with nails or screws. Once the tub is free of the walls, you will probably have to cut it into sections. One-piece tub-shower units will not normally fit through interior doors or travel up or down finished stairs. You can cut the unit into pieces with a hacksaw blade, but the job goes much more quickly with a reciprocating saw. Again, eye protection is strongly recommended.

Standard bathtubs (ones without integral shower walls) are not normally attached to the stud walls; they sit on supports. All that is required to remove a standard tub is to lift and slide it out of the opening. This sounds easier than it sometimes is.

Plastic, fiberglass, and steel bathtubs can be removed without too much strain, but cast-iron tubs are another matter entirely. Cast-iron tubs can weigh over 400 pounds and wrestling one out of its resting place can be very difficult, even for seasoned professionals. Many professionals use sledgehammers to break cast-iron tubs into manageable pieces. If you do this, wear eye and ear protection, along with clothing that will protect you from sharp pieces of the tub that may fly into your body.

Heating

You may not have to do much with the heating system. If the heat comes in through ducts in the floor, all you have to do is remove the register from the duct and protect the open duct from falling debris. You can stuff a towel in the duct or cut a piece of plywood to cover the opening.

When hot-water baseboard heat is in place, you will probably want to remove the baseboard heating element. This will require shutting down the boiler and may require draining the heating system. If you are working on the top floor of a home, you shouldn't need to drain much water from the heating system before cutting the supply and return pipes at the baseboard unit. However, if there is heat installed in rooms above the bathroom, drain the heating system to a point below the bathroom.

There will be removable end-caps on the baseboard heating unit. Remove these caps by pulling them off to reveal the supply and return pipes. The pipes should be copper. If they are, they can be cut with a hacksaw or a pipe cutter. Once the pipes are cut, remove the screws that hold the baseboard unit to the wall and remove the heating unit. None of this is difficult, but you must make sure that the heating system is shut down and that all water has been drained to a point below the pipes that you are cutting.

If you are working in an old house that is equipped with radiators, try to avoid removing them. Old radiators can become damaged when moved, and they are hard and expensive to replace. When conditions force you to remove a radiator, shut down the heating system and drain it to a point below the bathroom. Disconnect the piping from each side of the radiator. Be advised, ugly, black water is likely to come running out of the radiator either when it is disconnected or moved. Be very gentle with the radiator. Avoid banging it into a wall or slamming it down. Even minor movement can loosen fittings or the heating sections and set the stage for a leak when the heating unit is reinstalled. Store the unit standing up, don't lay it on its side. Cast-iron radiators can be very heavy, so make sure you are prepared to deal with the weight. You can do this with either extra human help or with the aid of a hand truck.

Electrical

Now that all the plumbing and heating is out of the way, you are ready to work with electrical devices and fixtures. Turn the power off to the bathroom. If you are going to remove electrical fixtures, be certain there is no electricity turned on to the wires. Use an electrical meter to test each wire before working with it. If you don't know how to use an electrical meter, you have no business working with electricity; call in an electrician. Never trust any wire. Always take a meter reading and always protect the bare ends of wires by covering them with wire nuts.

With the power turned off, remove all cover plates from switches and outlets. Remove the globes or shades on the lights and remove the light bulbs. Most electrical fixtures are attached to their electrical boxes with threaded rod and nuts. Remove these nuts and the fixture should come loose. Remove the wire nuts (plastic covers protecting the wires) and test for electricity. When you are sure the power is off, separate the fixture wires from the house wiring. Install wire nuts on the house wiring and tuck it back into the electrical box. Make sure that there is no exposed wiring. You can't be certain that someone might not turn the power back on at some point prior to the job's completion. Leaving any bare wiring exposed can result in electrocution or fire.

If you are dealing with electric baseboard heat, it should be attached to the wall with screws. Before handling the wiring to the heat, make sure the electricity is off. You cannot count on all of your bathroom wiring to be on the same circuit. Heating units usually have their own, separate circuit. Just because the light in the bathroom doesn't have power coming to it doesn't mean the heat is safe to work with.

Once you feel that the power is off to the heat, remove the cover of the baseboard unit. Remove the wire nuts from the wires and use your meter to test for electricity. If the wires don't show any current, unwind the wires and install wire nuts on the feed wires. Remove the screws from the back of the baseboard panel and lift it out of place. Use electrical tape to wrap the wire nuts and the feed wires to reduce the odds of injury to someone. You can't afford to have the wire nuts accidentally removed from the house wires in the event power is returned to the wires inadvertently.

Walls and ceilings

Removing finished walls and ceilings is not difficult if you are working with drywall. Use a hammer to open the walls and ceiling and to expose all wiring, plumbing, and heating. A dust mask will help protect you from the massive amounts of dust this process will create. You can then either continue to demo the walls and ceiling with a hammer, or you can cut the drywall out with a saw. Window and door trim will also have to be removed during this stage.

If the walls are made of plaster, you have a lot more work in front of you. A reciprocating saw is the fastest way to cut through plaster and the lathe behind it. However, use a hammer to open sections of the wall before running the saw through the plaster. It is easy to cut wires and plumbing by accident. The inspection holes made with a hammer will help you to avoid accidental cutting of unwanted materials.

Flooring

Removing vinyl flooring is not difficult. Start by removing all baseboard trim and shoe molding. When the molding is removed, the edges of the flooring will be exposed. You may be able to grasp the ends and pull the flooring up. If the floor is difficult to remove, you can use a floor scraper to remove it. It may be necessary to heat the floor in order to remove the vinyl flooring. This can be done with various types of heaters. If you don't own an appropriate heater, you should be able to rent one from a local rental store.

If you are removing a ceramic tile floor, you can chisel the tiles up or break them out with a hammer. Remember to protect your eyes and body from the sharp slivers created by breaking the tile. Also keep in mind that pounding the floor may damage finished ceilings below the bathroom.

Odds and ends

There will be some odds and ends to tend to. Go around the bathroom and remove all existing nails that protrude from the walls and ceiling. Sweep the floor and scrape it until it is clean. Cap all pipes to keep debris from entering them. Make sure all electrical wires are protected with wire nuts. Look around and tidy up any loose ends.

When you are involved with remodeling you are sure to run into some unexpected conditions. You might find that the subfloor or even the floor joists have been damaged by water leaks. When you open a wall you might come face to face with a nest of angry bees. There are all sorts of things that can disrupt your remodeling plans. Remodeling is work that you can plan for, but you can never be sure of exactly what you are getting yourself into until the work begins.

Alternative facilities

If you are ripping a bathroom out of a house where additional bathrooms are not present, you may have to provide alternative facilities for your customer and workers. A portable toilet that can be set somewhere on the property is one of the best options. If this is not feasible, see if the homeowners can make short-term arrangements with a neighbor. When all else fails, have your plumber reset the toilet on the stripped-out floor for temporary use. This will run your cost up a little, but people need sanitary waste provisions.

Once you establish a system for your demolition work, the job can go quickly and smoothly. Plan the work in advance and attempt to stick to your plan. Revise your methods from job to job as you gain experience. It will not take many rip-outs to get you to where the work is simple and efficient.

Selecting the best plumbing fixtures

SELECTING the best plumbing fixtures can be a confusing task. There are a lot of plumbing fixtures available to choose from. You and your customers can run the gamut from designer, one-piece toilets to special, replica toilets where the flush tank hangs high above the bowl and is operated with a pull-chain. Bidets are not common in the United States, but they are used in high-end housing and at other times. Bathtubs, whirlpool tubs, shower stalls, lavatories, and faucets are also part of the decision-making process when choosing the best plumbing fixtures. Your customers will probably enjoy the shopping process, but you may be called upon for advice. How much do you know about different types of plumbing fixtures? Will you be able to answer the questions your customers have, or will you have to stall them while you check with suppliers and plumbers? The more you know about fixtures before questions are asked, the more professional you will be as a general contractor.

What approach should you take in becoming educated on the subject of plumbing fixtures? There are different ways in which to expand your knowledge. Reading this chapter is one way of making yourself aware of options for your customers. Contacting plumbing suppliers and manufacturers is another way of gathering information. Your plumbing contractor is also a potential source of information. By using two or all three of these options, you can become effective in pointing your customers in the right direction.

You may be wondering right now what difference it makes about the style of a toilet. After all, a toilet is a toilet, right? Wrong! There are differences, and some of them are cosmetic and others are functional. Yes, the net effect of the desired result for a toilet is the same with any type, but there are clearly differences. Saying that all toilets are the same is similar to saying that all trucks are the same. Are you partial to a particular brand of truck? Do styling, color, amenities, and design come into play when you shop for a new vehicle? You are probably not very different from other consumers when it comes to making purchases. Most people buy products that they want, need, and like. A toilet fits all three elements of this profile. They are needed in modern life, and are wanted, and people have personal preferences in what makes a toilet appealing.

Who is going to care whether they bathe in a bathtub made of steel or a bathtub made of fiberglass? A lot of people. Why do people insist

▲

135

on paying more for heavy, cast-iron bathtubs when less-expensive and lighter versions of bathtubs are available? How important is a slip-resistant finish on the floor of a bathtub? It may not matter much to some people and be extremely important to others. Are tall people going to be comfortable in a standard, five-foot bathtub? Probably not as comfortable as they would be in a six-foot tub.

If a customer asks you to describe the advantages of a pedestal lavatory (Fig. 8-1) over a vanity, are you prepared to do so? Why are some pedestal lavatories very expensive while others are quite inexpensive? What are your feelings on steel lavatories when they are compared to china lavatories? Is there any reason not to use the old standby of a wall-hung lavatory? Are vanity tops with integral bowls better than tops with drop-in lavatory bowls? How are you doing so far on these questions?

Is there really a difference that is worth the price when you compare low-end fixtures to high-end fixtures (Fig. 8-2)? If so, what is it? Will your customers actually get more for their money if they invest in top-of-the-line brand names? Should customers choose on appearance, name, or design when it comes to selecting bathroom fixtures? We are just beginning to ask some key questions here, so you might see, already, that there can be a lot to learn about choosing plumbing

Figure 8-1

Here is a top-of-the-line pedestal lavatory. Courtesy of Porcher Ltd., a Division of American Standard, Inc.

Figure 8-2

Using high quality fixtures, like these, adds elegance to a bathroom. Courtesy of
St. Thomas Creations

fixtures. To expand on this, let's move through the various fixture
types on a one-by-one basis.

Toilets

Toilets are found in every modern bathroom (Fig. 8-3). Whether you
call it a water closet, a toilet, or a flush, it can be one of the most
important seats in the house. There is no argument that every house
should have a toilet, but which type of toilet is best? Would you
recommend a two-piece combination toilet, a one-piece toilet, a wall-
hung toilet, or some designer style? Obliviously, you need to know
something about the job and your customer's interest before you can
make such a recommendation. Two-piece combination toilets are the
most common type of residential toilets. It's rare to find a wall-hung
toilet in a home, but they do exist. One-piece toilets are popular in
up-scale homes, and designer toilets are found here and there. Based

Figure 8-3

One-piece toilets are very popular. Courtesy of American Chinaware

on this, the two-piece combination would seem to be the logical choice, and it probably is. However, you have to invest some time in talking with your customer to be sure of what you are recommending.

When you get on the subject of toilets, you are opening up a lot to talk about. For example, will a low-volume flush perform properly in an old home? Many codes require new toilets to flush with a minimal amount of water. This is okay in a new home, where smooth piping is installed on a proper grade. However, if the same toilet is installed in an old home, where rough cast-iron pipe is installed at a questionable grade, your customer could experience back-up problems in the drain line. This can be overcome with toilets that have help in their flushing, such as a vacuum-flush. The reasoning behind low-volume flushes is sound, but you may have to prepare your customers for potential problems if their existing piping is old and substandard.

Two-piece toilets

Two-piece toilets are the most common type of water closet found in an average home. These toilets are cost-effective, attractive enough, and quite functional. They can be purchased for less than $100, and they usually get their job done with dependability. This might be enough for you to recommend two-piece toilets as a standard.

However, even within the classification of two-piece toilets, there can be a lot of difference in appearance. The most noticeable difference, and the one I've received the most negative feedback on, is the outward appearance of the toilet's trap. Some toilets are made so that the shape of the trap is not visible. Other toilets are made in a way that the shape of the trap shows up on the closet bowl. Many customers don't like this look. Either type of toilet will function just fine, but the appearance of the exposed-trap toilet ruins it in the eyes of some customers. Yet other people don't see the look as being distasteful. Other considerations in appearance revolve around the overall design of a toilet. There usually isn't much difference between two-piece toilets. There can be some difference in the sculpturing of the tank top. You will find some closet-bolt covers that are rectangular and some that are round. The base of the toilet might be rounded or squared off. All of these are aesthetic issues, but aesthetics can be very important to a customer.

The height of a toilet bowl can be important to customers. Most toilets are similar in height. In addition to standard toilets, you can go to elevated closet bowls, usually intended for handicap use, that will not require as much effort from your customer in sitting down and standing up. A lot of customers don't consider themselves handicapped and would probably never explore this option. However, if you make your customers aware of this type of toilet, you might be surprised at how many people will find it appealing. Customers for higher closet bowls don't have to be physically restricted. For example, a tall person might find an elevated bowl to be more comfortable than a standard bowl. Remember, as a general contractor or remodeler, it is your job to offer your customers as many options as possible to ensure that they get the best job possible.

Much of the decision in choosing a toilet will boil down to customer preference. This involves everything from the general type of toilet to the color of the fixture. Will a toilet that is white in color work less efficiently than one that is in a designer or high-fashion color? Of course not, so why are the special colors so popular? People like to have bathrooms that are pleasant to the eye and that are different. Pull out the stops and show your customers all of their options when you are helping with a bathroom design.

Plumbing manufacturers will be happy to supply you with a wide array of product information. You can start gathering a collection of pictures and descriptions of plumbing fixtures from your plumbing contractor or plumbing supplier. If you want to go further in building your selection library for customers, you can contact manufacturers directly. Compile an extensive list of fixtures available to you. When you have shelves full of catalogs, you can help your customers make better decisions more quickly. Not only will this make your customers happy, but also it will separate you from most of your competitors, which is never a bad situation when you are bidding a job.

One-piece toilets

One-piece toilets are desirable to customers for many reasons. The profile of a one-piece toilet is typically lower than that of a combination toilet. One-piece toilets are a little easier to clean, and this appeals to some homeowners. There is a certain amount of prestige that goes along with owning a one-piece toilet. Does a one-piece toilet work better than a combination toilet? No, not really, but customers often prefer the one-piece product. The cost of a one-piece toilet is usually somewhat more than it would be for a combination toilet. This additional cost is justified by some customers in the designer look. There is a distinctive look to a one-piece toilet.

Deciding when to use a one-piece toilet is usually best left to the customer. However, there might be times when you would be doing your customer a favor to either recommend or discourage the use of such a toilet. If you are working in an exclusive neighborhood of homes where one-piece toilets are prevalent, then your customer should probably opt for the same type of toilet. On the other hand, if the home you are remodeling is not in an area known to support the cost of such fixtures, you might recommend that your customers consider a combination toilet. The final decision will rest on the shoulders of the customer, but you should make yourself available for advice whenever possible and feasible.

Wall-hung toilets

Wall-hung toilets are something of a rarity in modern bathrooms. There are houses where these types of toilets have been installed, but in all my years in the business I doubt if I've seen more than five of them in homes. However, I have had customers express an interest in wall-hung toilets because it is easy to clean the floor under them. This is, I suppose, a legitimate reason to consider a wall-hung fixture. All in all, I would recommend against a wall-hung toilet in a residential bathroom. If your customer is adamant about having a wall-hung toilet installed as a replacement to a floor-mounted toilet, your plumber is going to have to make some modifications in the existing piping. You, your carpenters, or your plumber will also have to install some serious support between wall studs to hold the weight of the toilet and its user.

Designer toilets

Designer toilets are not used often in standard homes. There are, however, some very nice toilets available for customers who are willing to pay the price for them. The one type of designer toilet that I've gotten the most requests for is a replica of an old-fashioned toilet. The toilet bowl sets on the floor and a long flush tube connects the bowl to a flush tank that hangs high on a wall. The flush is operated by pulling a chain that extends down from the flush tank. This is a very tasteful design that looks good, but it is also quite expensive. Few homes can justify the expense of such a toilet. It is a good idea, however, to develop a selection of customized toilets for your customers to choose from. More and more companies are catering to designer series, and you may find that you can increase your profits on jobs when you have literature on designer toilets to show your customers.

Bathtubs

Bathtubs are usually found in most bathrooms. There are occasions when bathtubs are waived in lieu of showers, but a majority of bathrooms incorporate the use of bathtubs and bath-shower combination units. Fiberglass tub-shower combinations are, by far, the most commonly used type of bathing fixture installed in modern bathrooms. One-piece units are used for new construction, and modular units are used in remodeling jobs. But there are still a fair number of cast-iron bathtubs being installed. Steel tubs are an inexpensive alternative to cast-iron, and acrylic tubs are a possible substitution for fiberglass models. Then there are designer series offered for bathtubs (Fig. 8-4). Which type of tub is right for your customer?

Figure 8-4

Claw-foot bathtubs are popular for both their look and their depth.
Courtesy of the Sanderson Garden Collection

Porcelain-covered bathtubs

Porcelain-covered bathtubs can be made of cast-iron or steel. The least expensive of the two is the steel tub. General appearance of the two types of tubs is very similar. Cast-iron tubs weigh much more than steel tubs, and most plumbers would be much happier carrying a steel tub upstairs. Cast-iron tubs are durable, but their finish can be chipped almost as easily as the finish on a steel tub. One complaint about steel tubs that are installed above living space is the pinging noise that can be heard when someone is running water in the tub. This sound problem is not as prolific with cast-iron. A steel tub can be purchased with a foam, sound-deadening block that will help to cut down on the pinging noise. The tub can also be set in sand to further reduce the noise. Even fiberglass batt insulation can be installed to quiet the tub.

Steel tubs are not as rugged as cast-iron tubs. On super-cheap models, your customers may feel the base of a steel tub bending or giving under their weight when standing in it. This is not usually the case with more expensive steel tubs that have better support struts installed beneath them. I have installed a lot of cast-iron and steel tubs. The cost and the lighter weight of steel tubs appeal to me. However, there are still customers who are willing to pay more for cast-iron, and this is an option you should offer your customers. Another benefit to cast-iron tubs is that they are usually available in a wider variety of colors, and this can be very important in specialized bathrooms. You might also want to look into replica fixtures that will bring back a taste of the past (Fig. 8-5).

One-piece tub-shower combinations

One-piece tub-shower combinations are the industry standard for new construction. The units are ideal in that they don't have seams or tile to leak, and they are easy to clean. A standard, one-piece tub-shower combination is very affordable and quite durable. They can be cracked, but they are more resistant to items being dropped in them than a porcelain-covered tub would be. Occupants of these tubs will find them warmer to sit in than a porcelain tub. Pinging is not a problem with fiberglass or acrylic tubs, and the slip-resistant flooring options are good. For all practical purposes, I cannot think of a better choice in a bathing facility. The drawback to this type of unit for remodeling is that it will not normally fit through a standard door opening, and it can rarely be carried up an existing stairway to an upstairs bathroom. To get around this, you have to go to a modular unit.

Figure 8-5

Replica fixtures are quite popular. Courtesy of Ackermann

Modular tub-shower combinations

It's difficult to tell some modular tub-shower units from one-piece units. High-quality modular units are excellent choices for remodeling jobs. Cheap modular units can cause problems. Inexpensive models can have defined seams that are not attractive and that may leak. I strongly suggest that you and your customers pay for the advantages of name-brand, high-quality modular units if such a unit is needed. Modular units range in the number of pieces from two to five sections. The key is finding a model that suits your customer's visual desire and that will install so that you will not be faced with call-back problems in the future. You must assess your ability to do a job before you commit to a specific type of modular unit. Two-piece units are good, if you can get them to the installation location. I've installed a number of modular units with multiple pieces and have never had a problem with any of them when they were installed and sealed properly.

Whirlpool tubs

Whirlpool tubs have gained popularity over the years (Fig. 8-6). There are units available that will fit into the space allotted for a standard bathtub. These small units are not spacious and they don't contain a lot of whirlpool jets, but they do offer whirlpool features in a compact package. Most customers will prefer whirlpools that are deeper and larger than a standard tub. This can cause some problems in small bathrooms where enlargement of the room is not feasible. Fiberglass is the material that most whirlpools are made of, but other types of materials are available.

The three keys, in my opinion, to a whirlpool are depth, the number of whirlpool jets, and the size of the motor that will power the unit (Fig. 8-7). The motor should be rated beyond a minimal size. This will give the motor a chance to last longer and to produce better whirlpool results. Depth is important, so that the user can submerge deeper into the swirling water. A whirlpool that is short and shallow limits a user's ability to gain full benefit from the jet action. Small whirlpool tubs may

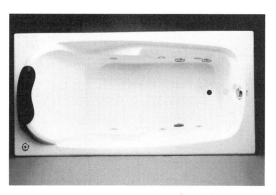

Figure 8-6

An acrylic whirlpool tub. Courtesy of Americh

Figure 8-7

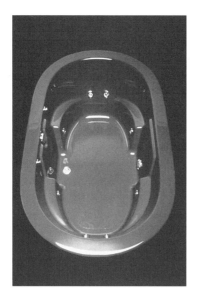

Unique tub shapes, such as this oval whirlpool, add to a new bathroom. Courtesy of Americh

have only four jets. This is enough for some people, but I feel that there should be at least six jets on most models.

Whirlpools range widely in cost. It's possible to get a decent unit for under $1,000, but most good models will range between $1,000 and $2,000, with larger units going for much more. A lot of research should be put into whirlpool units before a purchase is made. If a large unit is selected, you may have to strengthen the floor that will support the tub. The high volume of water held by four- and six-person whirlpools can produce a lot of weight. An alternative to a standard whirlpool is a soaking tub with whirlpool jets. These tubs are deep enough for a bather to submerge to neck level. Yet, the shape and design can accommodate a smaller space. Seats in garden tubs make soaking very comfortable, and a user gets full benefit from the whirlpool jets.

Showers

Showers are sometimes used in place of bathtubs, and they are often installed in addition to bathtubs. The last home I built for myself had a shower stall and a whirlpool tub in the master bathroom. This is a common and desirable concept. Many of today's showers are made of fiberglass or acrylic. These are both good choices, with fiberglass being less expensive. A standard shower stall can be purchased in many sizes, but a three-foot-by-three-foot size is, in my opinion, as small as you should go, whenever possible. A four-foot shower stall is even nicer. Many larger showers have seats molded right into the unit, and this is a nice touch. If space is tight, you can consider a corner shower. Your customer might prefer a shower base with tile walls. Stay away from the cheap metal shower stalls. They are not sturdy and they do rust afer years of use.

Lavatories

Lavatories offer a huge array of options for homeowners, remodelers, and builders. The two most common types are vanity tops with integral lavatory bowls and self-rimming, drop-in lavatory bowls (Fig. 8-8). Wall-hung lavatories (Fig. 8-9) used to be standard equipment, but few new bathrooms are using this time-tested design. Pedestal lavatories (Fig. 8-10) are popular in designer bathrooms, and vanities with tops (Fig. 8-11) are always appreciated for their storage capabilities. Many designer lavatories have evolved over the years, and they offer some decorative styling that you won't find in standard lavatories (Figs. 8-12–8-15). To expand on these options, let's look at some types of lavatories on a case-by-case study.

Figure 8-8

A designer drop-in lavatory. Courtesy of American Chinaware

Figure 8-9

A new, designer lavatory with an old look. Courtesy of Waterworks

Figure 8-10

A hand-painted pedestal lavatory. Courtesy of Porcher Ltd., a Division of American Standard, Inc.

Figure 8-11

A designer lavatory in a counter. Courtesy of Wilsonart

Wall-hung lavatories

Wall-hung lavatories are usually very inexpensive and quite functional. However, they are not in much demand today. People seem to want to get away from older standards and move ahead with newer designs. Wall-hung lavatories don't offer any storage space, and this is one reason why people tend to replace them with vanities and tops. If your customers are looking for an inexpensive, yet durable and functional lavatory, a wall-hung unit will do fine. Chances are, though, that your customers will not be inclined to install a wall-hung lavatory in a new or newly remodeled bathroom. There is now a modified

Figure 8-12

Painted lavatories can make a personal statement. Courtesy of American Chinaware

Figure 8-13

This is a hand basin that is affordable, but very special. Courtesy of Porcher Ltd., a Division of American Standard, Inc.

149

Figure 8-14

Pedestal lavatories are usually very popular. Courtesy of American Chinaware

Figure 8-15

A new type of wall-hung lavatory. Courtesy of Porcher Ltd., a Division of American Standard, Inc.

pedestal lavatory available (Fig. 8-16). Technically, this is a wall-hung unit.

Pedestal lavatories

Pedestal lavatories are basically dressed-up wall-hung lavatories. The pedestal that supports the wall-hung bowl hides all plumbing pipes and

Figure 8-16

Semi-pedestal lavatories are new, but are a good idea, especially when ADA regulations come into play. Courtesy of Absolute, a Division of American Standard, Inc.

adds a decorative image to a bathroom. Unfortunately, there is very little storage space with a pedestal lavatory. Some homeowners like this as a reason to reduce clutter. Others complain that there is no place to put anything. The major reason for using a pedestal lavatory is the look achieved with one. Prices on these units can be modest, but many models are expensive. In many cases, you could provide a vanity and a top for about the same amount of money that you'd pay for a pedestal lavatory.

One drawback to a pedestal lavatory might be in the case where small children use the bathroom. The pedestal of the lavatory can be knocked loose, and a rambunctious child might find this dangerous or amusing. Designer bathrooms often incorporate the use of a pedestal lavatory (Fig. 8-17), so this bathroom option should certainly be on your list of items to show your customers. The style (Fig. 8-18) and brand of the lavatory will be what forces the price to fluctuate.

Rimmed lavatories

Rimmed lavatories were used for years and are still used. Personally, I don't like the look of a rimmed lavatory. Many customers who are replacing this type of lavatory will probably tell you how difficult it is

Figure 8-17

This bathroom is very user-friendly, with two pedestal lavatories, a toilet, a bidet, and a bathing unit. Courtesy of Porcher Ltd., a Division of American Standard, Inc.

Figure 8-18

Here's a lavatory that is sure to attract attention. Courtesy of Porcher Ltd., a Division of American Standard, Inc.

to clean around the rim. Unless you have a customer who is determined to have a rimmed lavatory, I would avoid this type of fixture. A drop-in, self-rimming lavatory accomplishes the same goal with less installation trouble and easier maintenance.

Self-rimming lavatories

Self-rimming lavatories are very popular in certain regions. This style of lavatory sits in a counter top where a hole has been cut for it (Fig. 8-19). Installation is easy, and cleaning the fixture is not a problem. However, the seam between the lavatory bowl and the counter must be sealed well to prevent water leakage. Water accumulating on the counter can seep under the rim if the space is not sealed tightly with caulking. The expense of a self-rimming lavatory is minimal for a stock model, but a counter is required to mount the unit in. The counter usually sits atop a vanity, but some counters are free-standing. If you get into custom lavatories, the prices can shoot up quickly (Fig. 8-20 & 8-21).

Molded lavatory tops

Molded lavatory tops are ideal for use with vanities. These tops come in a wide variety of sizes and colors. The lavatory bowls are an integral part of the counter top, so leaking is not a problem. This type of unit is easy to clean, decorative, and very functional. Most of the tops are equipped with backsplash sections. For general use, I would recommend a vanity with a molded top and integral lavatory bowl.

Figure 8-19

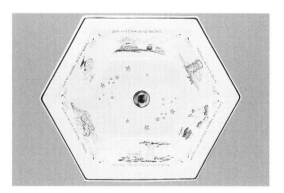

Handpainted basin are not common or inexpensive, but they are impressive. Courtesy of American Chinaware

Figure 8-20

Drop-in, handpainted lavatories are expensive, but can be worth the cost in special circumstances. Courtesy American Chinaware

Figure 8-21

Self-rimming lavatories are available in many colors. Courtesy of American Chinaware

Designer lavatories

Designer lavatories (Fig. 8-22) offer all that the human imagination can generate. There are lavatories that are wall hung with legs. You can get hand-painted lavatories that boast beautiful designs. The possibilities are nearly endless. This type of fixture is often expensive and no more functional than a builder-grade lavatory, but the look of a designer fixture can command a lot of attention. In special circumstances, this type of fixture is well worth considering.

Bidets

Bidets are interesting little fixtures that have not caught on well in the United States. They are, I understand, very popular in Europe. It is rare that I get a call to include this type of fixture in a bathroom, but I

Figure 8-22

Here you can see a semipedestal lavatory and a full pedestal lavatory.
Courtesy of Porcher Ltd., a Division of American Standard, Inc.

do occasionally get a call for one. They are not difficult to install, and they do make something of a conversational piece out of a bathroom. Based on my experience, you will not have customers beating down your door to install bidets, but you would still be wise to stock your book shelves with some photos and literature on them. There are many good brands and designs of bidets available, and you might find that you become known as a trend setter if you start installing some bidets on your jobs.

Faucets

Faucets are a necessary part of any bathroom. When you open a discussion on faucets, you can be talking for a long time. There is the simple debate of single-handle faucets (Fig. 8-23) versus two-handle faucets (Fig. 8-24), but the topic goes on and on. Should you install

Figure 8-23

A designer single-handle faucet. Courtesy of Hansa America

Figure 8-24

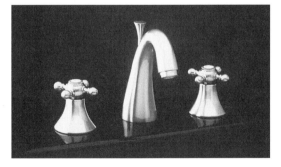

A nice two-handle faucet. Courtesy of Monet Faucets Ltd.

chrome, polished brass, antique brass, or gold (Fig. 8-25)? Gold is a profitable choice for you and your plumber, but gold faucets are too expensive for most installations. Chrome is the most common finish for a faucet, but the various tones of brass also attract their fair share of attention. Appearance is an important aspect of a faucet, but so is the practical application of the device. Let me expound on this a bit more.

Figure 8-25

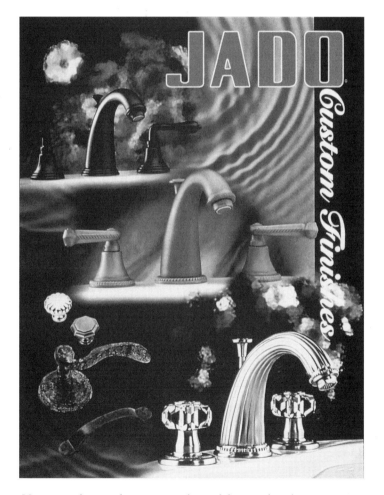

You can choose from a number of faucet finishes. Courtesy of JADO

Single-handle faucets

Single-handle faucets (Fig. 8-26) seem to be more popular than traditional two-handle faucets. Part of this, I believe, is appearance, and part of it is convenience in use. Personally, I like single-handle faucets. Many people, however, like two-handle faucets (Fig. 8-27). When young children or elderly people will be using faucets, I can see building a case in favor of two-handle faucets. There is less risk of injury from hot water when two-handle faucets are used, simply because the user will know which handle controls the hot and cold water. The choice between a two-handle or single-handle faucet is up to the customer, but you should have a good collection of information on both types to share with your customers.

Figure 8-26

This single-handle faucet offers the advantage of temperature limit stop.
Courtesy of Moen

Figure 8-27

Widespread faucets, like this one, work well with most pedestal lavatories.
Courtesy of Moen

Pressure-balanced faucets

Pressure-balanced faucets are required by code in many areas for faucets being installed in showers and bathtubs. Even if these devices are not required in your area, they are a good investment. It only takes a matter of seconds for a person to receive severe burns from super-hot water. A child who is taking a shower when a toilet is flushed or a washing machine cuts on could be burned badly if pressure-balanced or temperature-balanced faucets are not installed. This type of faucet costs a little more, but the price is well worth it.

The finish

The finish on a faucet has nothing to do with the performance of the device. Color or finish is strictly a matter of taste for the customer (Figs. 8-28 & 8-29). Chrome is a standard finish. Polished and antique brass are both used frequently. More exotic finishes, like gold, are used in expensive designer bathrooms. Most customers will probably be happy with a chrome faucet (Fig. 8-30), but make sure you ask before you place a bid, because the finish on a faucet can have a dramatic effect on the overall price.

Figure 8-28

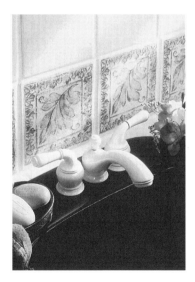

Colored faucets are gaining popularity. Courtesy of Moen

159

Figure 8-29

Polished brass faucets are frequently used in nice bathrooms. Courtesy of Moen

Figure 8-30

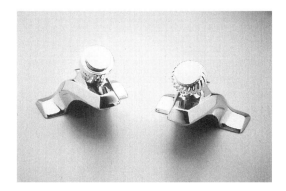

Chrome faucets represent the most common faucet finish. Courtesy of Moen

Handle selection

Handle selection may not seem like much of a big deal, but it can be. Some people have trouble with certain styles of handles. For example, a person with arthritis will normally find a blade-type handle (Fig. 8-31) much easier to use than a round handle. This doesn't mean that you must use institutional-type faucets. There are plenty of attractive faucets that offer bar or blade handles (Figs. 8-32 & 8-33) Single-handle faucets that use a lift-and-turn lever are also good for people with stiff fingers (Fig. 8-34). You should take this type of issue into consideration when offering your customers choices in handle designs.

Figure 8-31

Blade-type handles can be found on attractive faucets. Courtesy of Moen

Figure 8-32

This type of faucet is known as a mini-widespread design. Courtesy of Moen

Figure 8-33

Brushed nickel is an attractive finish for a faucet. Courtesy of JADO

Figure 8-34

Single-handle faucets are easy for almost anyone to use. Courtesy of Moen

Handle construction usually isn't as important as the design. Acrylic handles are popular, as are metal handles. Some faucets use wood inserts for eye appeal. I have had customers who did not like acrylic handles because they can be difficult to clean. On the whole though, I find that more customers like the cut-glass look of acrylic than dislike it. Many faucet manufacturers offer a number of different handle options for their faucets (Fig. 8-35). In other words, you might install a faucet with one type of handle and then change the handles later, to suit your customer's changing desires. This is not a bad option to consider when choosing a faucet.

Figure 8-35

Here is a faucet with polished brass lever handles. Courtesy of Moen

Replacing a two- or three-handle tub or tub-and-shower faucet with a single-handle faucet can be done without making a mess of existing surrounding walls for the bathing unit. Special wall plates—these remodeling plates allow you to install a single-handle faucet in place of a multihandle faucet without having to do major work to the finished wall. I've used the cover plates often, and they work very well. If you are remodeling a bathroom where the finished tub or shower surround will not be disturbed, this is a good way to upgrade to a single-handle faucet.

When you begin stocking your shelves with brochures on faucets you can accumulate quite a mass of material. Since faucets are one of the finishing touches in a bathroom, you should not limit the options of your customers. Talk to your plumbing contractor, your plumbing supplier, and plumbing manufacturers to get photographs and descriptions of the many faucets available to you and your customers.

Handicap fixtures

Handicap fixtures are available in many forms. You can offer your customers showers that will accept a wheelchair. Grab bars (Fig. 8-36) are useful in making a bathroom easier to use and safer. Nonslip surfaces in bathing units are a good idea for anyone. Toilets that sit high above the floor can be much easier for some people to use, even if the people don't have apparent physical disabilities. Blade-type handles on faucets are advantageous for some customers, as are bar-type, lift-and-turn handles. A counter with a drop-in lavatory or a wall-hung lavatory is a good choice if a bathroom user will be confined to a wheelchair. Personal shower attachments are also good for some people. Seats for bathtubs can be helpful, and a wider floor space might be necessary to meet the needs of some customers.

When you and your customers are choosing bathroom designs and fixtures, you should be actively involved in the process. Build a good reserve of literature for your customers to look through. Become familiar with the options available to your customers. By putting extra effort into your ability to communicate with your customers and to advise them, you set yourself apart from many of your competitors. This should result in your winning more bids and having happier customers.

Figure 8-36

Chapter 8

HM-360-32

HM-360-24

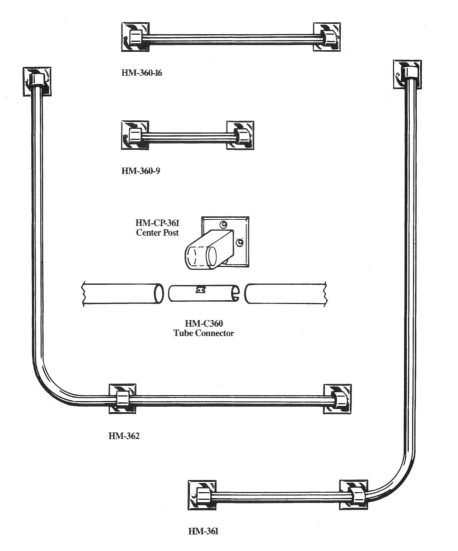

HM-360-16

HM-360-9

HM-CP-361
Center Post

HM-C360
Tube Connector

HM-362

HM-361

A selection of functional grab bars. Courtesy of Nutone Inc.

Plumbing points to ponder

THE CHANCES are that you will have a plumbing contractor doing your plumbing installations, but there are some plumbing points to ponder for the remodeler and general contractor. Knowing how high a pipe should rough in and how a basic installation should go is helpful when you are overseeing the work of a plumber. It's not necessary for you to be able to solder a copper joint or to install a toilet on your own, but it certainly doesn't hurt to possess some of these skills, in case you are caught in a tough spot and can't locate your plumber. Being able to assess existing conditions is also important. If you can look at a pipe and spot trouble before you bid a job, you're in a much stronger position. This chapter is going to prepare you for what to expect when a plumber starts to work on your next job.

Plumbing pipes used for drains and vents

There are many types of plumbing pipes that may be used for drains and vents. Most modern plumbing is done with plastic pipe, but older homes may be plumbed with pipes made from cast iron, galvanized steel, brass, lead, or some other type. Some of these pipes do not perform well once they age. If you open walls during a remodeling job, it may pay to replace sections of the existing plumbing to avoid future problems. Let's look at the most common types of pipe found in remodeling jobs to see their strong points and weaknesses.

Cast-iron pipe

Cast-iron pipe can be found in houses of all ages. It has not been used in residential plumbing extensively since the mid-1970s, but cast iron is still used today. If a home is more than twenty years old, there is a good chance it may have cast-iron drains and vents.

Cast-iron was typically used for large drains and vents. Galvanized steel pipe was often used in conjunction with cast-iron for small drains, like those of kitchen sinks, bathtubs, and lavatories. It is the galvanized steel pipe that you will be most likely to work with in simple remodeling jobs. And galvanized drains are good targets for removal, due to their tendency to close up and create stoppages.

Cast-iron joints used to be made with oakum and molten lead, and they still are today. However, technology has delivered special rubber adapters for making connections with cast iron in modern installations. There are three basic types of rubber adapters. One type resembles a doughnut and is placed in the hub of one pipe so that the end of another pipe can be inserted, making a watertight joint. The other two types are used with cast-iron pipe that does not have hubs. These adapters slide over the ends of two pipes and are held in place with stainless-steel clamps. Not only is this type of connection much easier to make, but it is also safer than working with hot lead.

Unless you are relocating a toilet or altering the main drainage and vent system in a home, it is unlikely you will have to work with cast-iron pipe. But since you may wish to relocate a toilet or do some other type of plumbing remodeling, let's take a quick look at how you can simplify the task of working with cast-iron pipe.

If you plan to cut cast-iron pipe, go to a tool rental place and rent a rachet-type soil-pipe cutter. These tools make quick, easy work of cutting cast iron. Basically all you have to do is wrap a special cutting chain around the pipe, secure the cutter, pump the handle a few times, and the pipe is cut cleanly. This is much easier than laboring your way through the pipe with a hacksaw. Cutting vertical sections of cast-iron pipe can be dangerous. If the pipe is not supported properly, the vertical piping could come crashing down on you. Before you cut a vertical pipe, make sure it is supported in a way to protect you. If you want to convert a piece of cast-iron pipe to another type of pipe, like plastic pipe, use a universal rubber adapter for the conversion. This will make the job fast and simple.

Galvanized steel pipe

If you find cast-iron pipe in a home, you will probably find some galvanized steel pipe, too. This pipe tends to rust and build up blockages in itself over the years. If you have the opportunity to replace galvanized pipe with plastic pipe, do it. You will be saving yourself from trouble that is likely to come in the future. Galvanized pipe can be cut with a hacksaw, and the same rubber adapters used to join cast-iron and plastic pipe can be used on galvanized pipe.

DWV copper

DWV copper was a popular drain and vent pipe for many years, and it is still found in many older homes. On the whole, copper drains and vents give very good service and should not need to be replaced. Copper drains and vents can be cut with a hacksaw, and the same universal adapters used with cast-iron and galvanized pipe can be used to convert copper to plastic.

Schedule-forty plastic pipe

Schedule-forty plastic pipe is the drain and vent pipe most often used in modern plumbing systems. There are two types of schedule-forty plastic pipe used in homes. They are ABS and PVC; ABS is a black and PVC is white. Both of these pipes are easy to work with, and either of them can be cut with a hacksaw or a standard carpenter's saw. Joints for these pipes are normally made with a solvent or glue. Most plumbing codes recommend a cleaner be used on plastic pipe and require that a primer be applied prior to gluing a joint. With universal rubber adapters, these pipes can be joined to any of the other types of drains and vents mentioned.

Plumbing pipes for potable water

Just as there are a number of approved drain and vent materials, there are also plenty of types of plumbing pipes for potable (drinking) water. Let's take a quick look at some of them.

Copper

Copper water pipe and tubing is found in more homes than any other type of water distribution pipe. It is a dependable material that provides years of service. Copper can be cut with a hacksaw, but roller-cutters will cut the pipe much smoother. The joints for copper pipe and tubing are usually made by soldering. This can be a problem for people who are not plumbers. Learning to solder watertight joints takes a little time and experience. One way to avoid soldering is to use compression fittings.

Compression fittings are available in all shapes and sizes. They are easy to install and they normally don't leak. If your joints are going to be concealed in a wall, compression fittings may not be a good idea, but they work great under sinks and in other accessible areas. There is some risk that the fittings will develop leaks as the pipes are vibrated with use. The leaks will be small and can be seen and corrected easily if they are visible. If, however, they are concealed in a wall, a small leak could go on for a long time, causing serious damage to building components, before the leak is detected.

CPVC pipe

CPVC pipe is another alternative for people lacking soldering skills. CPVC is a rigid plastic pipe that is put together with solvent joints. A cleaner and primer should be used on the pipe and fittings prior to gluing joints. CPVC can be cut quite easily with a hacksaw, and it is simple to install. You should, however, allow plenty of time for joints to dry before moving the pipe. If a fresh joint is bumped or twisted before the glue has dried, a leak is likely.

PEX pipe

PEX pipe is the new kid on the block. It is a flexible plastic pipe that can be installed much like electrical wiring. The pipe can be snaked through studs, and its flexibility allows for minimum joints. If you opt for PEX pipe, you will need to rent a special crimping tool or use special compression fittings. Do not attempt to make joints with standard stainless-steel clamps. PEX joints require the use of insert fittings and special crimp rings, or compression fittings made expressly for the pipe. Many professional plumbers feel that PEX pipe will eventually be more common than copper for potable water systems.

Where should you put the pipes?

Where should you put the pipes for various fixtures? The locations for pipes will vary with the type of fixtures being plumbed. Your plumbing supplier should be able to provide you with a rough-in book. The rough-in book will tell you exactly where to place pipes. Exact rough-in measurements are usually not critical, but they can be, especially with fixtures like pedestal lavatories. While it is impossible to predict exactly where pipes should go without rough-in specifications, there are some rule-of-thumb figures that will normally work. Let's take a fixture-by-fixture look at where you might want to put pipes.

Lavatories

Drains for lavatories should come out of the wall about seventeen inches above the subfloor. The center of the drain should line up with the center of the lavatory, approximately. For example, if you were installing a vanity against a side wall and the center of the lavatory would be fifteen inches away from the side wall, the drain should also be about fifteen inches from the side wall.

If the drain is roughed-in too low or too far to the left or right, the problem can be corrected. A tailpiece extension will compensate for a low trap, and fittings can be used to bring the trap arm closer to the fixture trap. But if you are roughing in pipes for a pedestal lavatory or a vanity that has drawers in it, precise location of the pipe becomes much more important. However, if the drain is roughed-in too high, you've got a problem that is not so easily corrected. The only solution to this problem is to rework the rough plumbing completely. When you inspect the work of your plumbing contractor, measure the rough-in to make sure it will not be too high. It's best to catch such mistakes as early as possible.

Water pipes for lavatories, if they come out of a wall, should be about twenty-one inches above the subfloor. Water pipes that come up out of the floor can be extended as needed. Hot water pipes should always be installed on the left side (as you face the lavatory to use it) of the lavatory.

Most lavatories have faucets with four-inch centers. This simply means that there are four inches between the hot and cold water supplies when measured from the center of one supply tube to the other. If the drain is roughed-in under the center of the lavatory, each water pipe will be about two inches from the center of the drain, on each side.

Toilets

Unless you are relocating a toilet, you will have no need to rough-in a new drain. However, if you are installing a new closet flange (the part a toilet sits on), the center of the flange should measure twelve and a half inches out from the back wall of where the toilet will set.
This measurement is based on measuring from a stud wall. If you are measuring from a finished wall, the distance would be an even twelve inches. It is possible to get toilets with different rough-in dimensions, such as ten inches and fourteen inches. Toilets like these cost more, but they can be a bargain when compared with moving an existing rough-in. You should be able to measure fifteen inches from the center of the closet flange to either side without hitting a wall or other fixture. Toilets are required to have a minimum free width of thirty inches for a proper installation.

Water supplies for toilets should be installed six inches above the subfloor, and they should be six inches to the left of the center of the drain.

Bathtubs

Drains for bathtubs are typically located fifteen inches off the stud wall where the back of the tub will rest. The drain is normally about four inches off the head wall, where the faucets will go. If you are installing a new tub drain, it is necessary to cut a hole in the subfloor for the drain before the tub is set in place permanently. The hole should extend from the head wall to a point about twelve inches away, and it should be about eight inches wide. This gives you a hole that is eight inches wide and twelve inches long. You will need most of this space to connect a tub waste and overflow to the tub. Bathtub faucets should be installed about twelve inches above the flood-level rim of the tub. The flood-level rim is the arm rest or location where water will first spill over the tub. Tub spouts are normally mounted about four inches above the flood-level rim of the tub. When installing the faucet for a tub-shower combination, the shower head outlet should be set about six-and-a-half feet above the subfloor.

Showers

A shower should not be installed permanently until a hole has been cut in the subfloor for the shower drain. Most shower drains are located in the center of the shower, but there are many types of showers where this is not the case. Consult a rough-in book or measure the actual drain location to determine where to rough in the shower drain. Faucets for showers should be installed about four feet above the subfloor. The shower head outlet should be installed about six-and-a-half feet above the subfloor.

Setting plumbing fixtures

When you are ready to set plumbing fixtures, the end of your job is in sight. Some plumbing fixtures, like toilets, must be handled with care, but the installation of most plumbing fixtures is not very difficult. Let's see what is involved with the installation of common plumbing fixtures.

Toilets

Toilets sit on and are bolted to closet flanges. Unless you have relocated the drain for a toilet, the existing closet flange should be able to be used. If you must install a new closet flange, install it so that the slots in the flange will allow the closet bolts to sit on either side of the center of the drain. The top of the flange should be flush with the finish flooring. If it is only slightly above the flooring you should not have any problems, but if it sits too high above the floor, the toilet will not mount properly.

Place closet bolts in the grooves of the flange and line them up with the center of the drain. Then install a wax ring over the drain opening in the flange. Now set the toilet bowl on the wax and press down firmly. The closet bolts should come up through the mounting holes in the base of the toilet.

Measure the distance from the back wall to the holes in the toilet where the seat will be installed. The two holes should be an equal distance from the back wall. If they are not, adjust the bowl until the holes are the same distance from the back wall.

Install the flat plastic caps (that came with the toilet) over the closet bolts. If metal washers were packed with the closet bolts, install them next. Install the closet-bolt nuts, and tighten them carefully: too much stress will break the bowl. Snap the plastic cover caps (that came with the toilet) over the bolts and onto the flat plastic disks you installed over the bolts. If the bolts are too long to allow the caps to seat, cut the bolts off with a hacksaw.

Uncrate the toilet tank and install the large sponge washer over the threaded piece that extends from the bottom of the tank. Then install the tank-to-bowl bolts. To do this, slide the heavy black washers over the bolts until they reach the heads of the bolts. Push the bolts through the toilet tank.

Pick the tank up and set it in place on the bowl. The sponge gasket and bolts should line up with the holes in the bowl. With the tank in place, slide metal washers over the tank-to-bowl bolts from beneath the bowl. Follow the washers with nuts and tighten them. Again, be careful; too much stress will crack the tank. Alternate between bolts as you tighten them. This allows the pressure to be applied evenly, reducing the chance of breakage. Tighten the bolts until the tank will not twist, and turn on the bowl. With this work done, you are ready to connect the water supply.

Turn the water to the toilet's supply pipe off. Cut the supply pipe off about three-quarters of an inch above the floor or past the wall, depending on where the pipe is coming from. Slide an escutcheon over the pipe, and install a cut-off valve. Compression valves require the least amount of skill and effort to install.

You will install a closet supply between the cut-off valve and the ballcock (the threads protruding past the bottom of the tank, in the left, front corner). Flexible supplies are the easiest type to install. Remove the ballcock nut from the threads at the bottom of the toilet tank. Hold a closet supply in place, and after measuring it, cut it to a suitable length.

Slide the ballcock nut onto the supply tube, with the threads facing the toilet tank. Slide the small nut from the cut-off valve onto the supply tube and follow it with the compression ferrule. Hold the supply up to the ballcock and run the ballcock nut up handtight. Insert the other end of the supply into the cut-off valve. Slide the ferrule and compression nut down to the threads on the cut-off, and tighten the nut. Then tighten the large ballcock nut.

Toilet seats generally have built-in bolts that fit through holes in the bowl. Put the seat in place and tighten the nuts that hold it in place. Some seats don't have attached bolts, but this simply means installing the provided bolts through existing holes and tightening them. This completes the toilet installation. After turning the water on, flush the toilet several times and check all connections to make sure none are leaking. When you are checking up on the work of your plumbing contractor, you should flush the toilet many times to test it for leaks. Wipe down all water connections with toilet tissue to expose leaks that are not visible.

Installing a lavatory

How you install a lavatory will depend on the type of lavatory you are using. Drop-in lavatories require a hole to be cut in the counter where they will be mounted. The bowls are set in the hole and are held in place by their weight and plumbing connections. A bead of caulking should be placed around the hole, on the surface of the counter, before these lavatories are set in place.

Rimmed lavatories also require a hole to be cut in the countertop, but they are not as easy to install as drop-in lavatories. Rimmed lavatories have a metal rim that is placed in the hole of the countertop. The lavatory bowl is then held up to the ring, from below, and secured with special clips.

Wall-hung lavatories hang on wall brackets. Wood backing must be installed during the rough-in phase so that there will be a firm surface to bolt the wall bracket to. Once the wall bracket is mounted, most wall-hung lavatories just sit on the bracket. A few types have additional mounting holes where lag bolts can be used to provide additional security that the bowl will not be knocked off the wall bracket. Most wall-hung lavatories (Fig. 9-1) are made to accept legs, but the legs are optional.

Figure 9-1

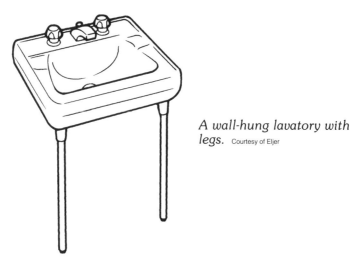

A wall-hung lavatory with legs. Courtesy of Eljer

Vanity tops with the lavatory bowl built into the top are the easiest to install. These tops simply sit on a vanity cabinet. Normally the tops are heavy enough to sit in place without any special attachments being required.

Figure 9-2

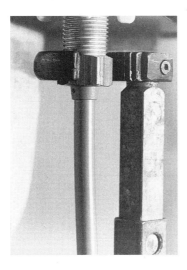

A basin wrench. Photo by Images International, a division of Lone Wolf Enterprises, Ltd.

Lavatory faucets and drains

All standard lavatory faucets and drains go together about the same. This job will be easier if you mount the faucet and drain assembly before you install the lavatory. A basin wrench (Fig. 9-2) may be necessary when working with faucets; they can be purchased inexpensively at any store that sells a selection of tools.

Many faucets come with gaskets that fit between the base of the faucet and the lavatory. If your faucet doesn't have one of these gaskets, make a gasket from plumber's putty. Roll the putty into a long round line and place it around the perimeter of the faucet base. Place the faucet on the lavatory with the threaded fittings going through the holes. Slide the ridged washers over the threaded fittings of the faucet, and then screw on the mounting nuts. Tighten these nuts until the faucet is firmly seated.

Lavatory supply tubes mount on the ends of the threaded fittings protruding below the lavatory. Slide supply nuts up the supply tubes and screw them onto the threaded fittings. The beveled heads of lavatory supplies prevent leaks. Now you are ready to connect the drainage. The first step is the assembly and installation of the pop-up drain assembly. Detailed instructions for the proper installation of the pop-up assembly should be packed with your faucet; read and follow the instructions provided by the manufacturer.

A pop-up assembly mounts in the hole in the bottom of a lavatory (Fig. 9-3). Unscrew the round trim piece from the shaft of the pop-up. This is the piece you are accustomed to seeing when you look into a lavatory. Roll up some plumber's putty and place a ring of it around the bottom of the trim piece. Slide the fat, tapered, black washer, which is on the threaded portion of the assembly, down on the threaded shaft. You may have to loosen the big nut that is on the threads to get the metal washer and rubber washer to move down on the assembly.

Apply pipe dope or a sealant tape to the threads of the pop-up assembly. With your hand under the lavatory bowl, push the threaded assembly up through the drainage hole. Screw the small trim piece, the one with the putty on it, onto the threads. Push the tapered gasket up to the bottom of the lavatory. Tighten the mounting nut until it pushes the metal washer up to the rubber washer and compresses the rubber washer. You should notice putty being squeezed out from under the trim ring as you tighten the nut.

When the mounting nut is tight, the metal pop-up rod that extends from the assembly should be pointing to the rear of the lavatory bowl. Take the thin metal rod, the rod used to open and close the lavatory drain, and push it through the small hole in the center of the faucet.

You should see a thin metal clip on the end of the rod that extends from the pop-up assembly. Remove the first edge of this clip from the round rod. Take the perforated metal strip that was packed with the pop-up and slide it over the pop-up rod. You can use any of the holes for starters. Now slide the edge of the thin metal clip back onto the pop-up rod; this will hold the perforated strip in place.

At the other end of the perforated strip there will be a hole and a setscrew. Loosen the setscrew and slide the pop-up rod, the rod used to open and close the drain, through the hole. Hold the rod so that about one-and-a-half inches of it is protruding above the top of the faucet. Tighten the setscrew. Pull up on the pop-up rod and see that it

Figure 9-3

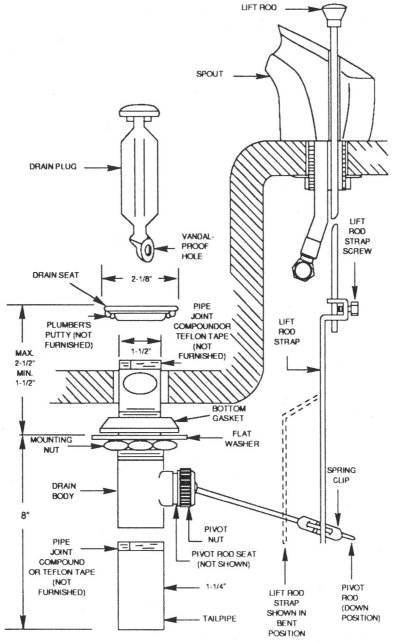

LIFT ROD

SPOUT

DRAIN PLUG

VANDAL-PROOF HOLE

DRAIN SEAT

2-1/8"

PIPE JOINT COMPOUND OR TEFLON TAPE (NOT FURNISHED)

PLUMBER'S PUTTY (NOT FURNISHED)

MAX. 2-1/2" MIN. 1-1/2"

1-1/2"

LIFT ROD STRAP

LIFT ROD STRAP SCREW

BOTTOM GASKET

MOUNTING NUT

FLAT WASHER

DRAIN BODY

8"

SPRING CLIP

PIVOT NUT

PIPE JOINT COMPOUND OR TEFLON TAPE (NOT FURNISHED)

PIVOT ROD SEAT (NOT SHOWN)

1-1/4"

LIFT ROD STRAP SHOWN IN BENT POSITION

PIVOT ROD (DOWN POSITION)

TAILPIPE

Detail of a pop-up assembly for a lavatory. Courtesy of Moen

179

operates the pop-up plug. The pop-up plug is the stopper in the sink drain. You can test this best after all connections are made to the water and drainage systems.

There should be a one-and-one-fourth-inch chrome tailpiece (round tubular piece) packed with the pop-up assembly. The tailpiece will have fine threads on one end and no threads on the other. Coat the threads with pipe dope of sealant tape. Screw the tailpiece into the bottom of the pop-up assembly.

Now you are ready to install the trap. First, slide an escutcheon over the trap arm (the pipe coming out of the wall). Lavatory traps are normally one-and-one-fourth inches; however, you can use a one-and-one-half-inch trap with a reducing nut on the end that connects to the tailpiece. Assuming you used plastic pipe for your rough-in, you may either glue your trap directly to the trap arm, if you are using a schedule-40 trap, or you may use a trap adapter. A trap adapter will be needed if the trap is metal and the trap arm is plastic. Trap adapters glue onto pipe, just like any other fitting. One end of the adapter is equipped with threads, to accept a slip-nut.

Start by placing the trap on the tailpiece. To do this, remove the slip-nut from the vertical section of the trap. Slide the slip-nut onto the tailpiece and follow it with the washer that was under it; the washer may be nylon or rubber. Put the trap on the tailpiece, and check the alignment with the trap arm. It may be necessary to use a fitting to offset the trap arm in the direction of the trap.

If the trap is below the trap arm, you will have to shorten the tailpiece. The tailpiece is best cut with a pair of roller-cutters, but it can be cut with a hacksaw. You may have to remove the tailpiece to cut it. If the trap is too high, you can use a tailpiece extension to lower it. A tailpiece extension is a tubular section that fits between the trap and the tailpiece. The extension may be plastic or metal, and it is held in place with slip-nuts and washers.

Once the trap is at the proper height, you must determine if the trap arm needs to be cut or extended. Extending the trap arm can be done with a regular coupling and pipe section. If you are using a schedule-40 plastic trap, it is glued onto the trap arm. If you are using a metal trap, the long section of the trap will slip into a trap adapter. You zmay have to shorten the length of the trap's horizontal section. When using a trap adapter, slide the slip-nut and washer on the trap section; then insert the trap section into the adapter and secure it by tightening the slip-nut. Once the trap-to-trap-arm connection is complete, tighten the slip-nut at the tailpiece.

With the water cut off to the supply pipes, install the cut-offs for the lavatory. Remove the aerator (the piece screwed onto the faucet spout) from the faucet. If you don't remove the aerator before you run water through the faucet for the first time, it will often become blocked with debris and cause an erratic water stream. Other than to test for leaks, your work with the lavatory is done.

Installing tub and shower trim

Installing tub and shower trim is easy (Fig. 9-4), but connecting a tub waste and overflow is difficult if you don't have any help. Let's take a look at what you need to know to trim out your tub or shower. When

Rough-In Measurements

Figure 9-4

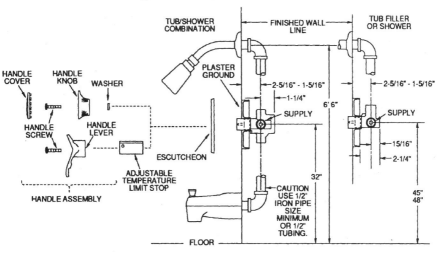

Rough-in measurements for tub-shower faucet. Courtesy of Moen

you want to install shower trim, start with the shower head. Be sure the main water supply is turned off, and unscrew the stub-out from the shower-head ell. Slide the escutcheon that came with the shower assembly over the shower arm. Apply pipe dope or sealant tape to the threads on each end of the shower arm. Screw the shower head on the short section of the arm, where the bend is. Screw the long section of the arm into the threaded ell in the wall.

Use an adjustable wrench, on the flats around the shower head, to tighten all connections. If you must use pliers on the arm, keep them close to the wall so that the escutcheon will hide scratch marks. Now you are ready to trim out the shower valve. How this is done will depend on the type of faucet you roughed-in. Follow manufacturer's suggestions. If you installed a single-handle unit, you will normally install a large escutcheon first. These escutcheons normally incorporate the use of a foam gasket, eliminating the need for plumber's putty. Then the handle is installed and the cover cap is snapped into place over the handle screw.

If you are using a two-handle faucet, you will normally screw chrome collars over the faucet stems. These may be followed by escutcheons, or the escutcheons may be an integral part of the sleeves. Putty should be placed where the escutcheons come into contact with the tub wall. Then the handles are installed.

Tub faucets are trimmed out in the same ways as shower faucets. However, you will have a tub spout to install. Some tub spouts slide over a piece of copper tubing and are held in place with a setscrew. Many tub spouts have female-threaded connections either at the inlet or the outlet of the spout. If you are dealing with a threaded connection, you must solder a male adapter onto the stub-out from your tub valve or use a threaded ell and galvanized nipple. The type of spout that slides over the copper and attaches with a setscrew is by far the easiest to install. You should place plumber's putty on the tub spout where it comes into contact with the tub wall.

Tub wastes are difficult to install when you are working alone. The tub waste and overflow can take several forms. It may be made of metal or plastic. It can use a trip lever, a push button, a twist-and-turn stopper, or an old-fashioned rubber stopper. The tub waste may go together with slip-nuts or glued joints. Follow the directions that come with your tub waste.

The first step for installing a tub waste is the mounting of the drain. Unscrew the chrome drain from the tub shoe. You will see a thick black washer. Install a ring of putty around the chrome drain, and apply pipe dope to the threads. Hold the tub shoe under the tub so that it lines up with the drain hole. Screw the chrome drain into the female threads of the shoe. The black washer should be on the bottom of the tub, between the tub and the shoe. Once the chrome drain is hand tight, leave it alone, for now.

The tub shoe has a tubular drainage pipe extending from it. Make this drain point towards the head of the tub, where the faucets are. Take the tee that came with the tub waste and put it on the drainage tube from the shoe. The long drainage tube that will accept the tub's overflow should be placed in the top of the tee. You want the face of the overflow tube to line up with the overflow hole in the tub. Cut the tubing on the overflow or shoe as needed for a proper fit. The cuts are best made with roller-cutters, but they can be made with a hack saw.

You should have a sponge gasket in your assortment of parts. This gasket will be placed on the face of the overflow tubing, between the back of the tub and the overflow head. From inside the bathtub, install the face plate for the overflow. For trip-lever styles you will have to fish the trip mechanism down the overflow tubing. For other types of tub wastes you will only have a cover plate to screw on. Tighten the screws until the sponge gasket is compressed.

Now tighten the drain. This can be done by crossing two large screwdrivers and using them between the crossbars of the drain. Turn the drain clockwise until the putty spreads out from under the drain. The last step is connecting the tub waste to the trap. This can be done with trap adapters or glue joints, depending upon the type of tub waste you have used. Apply joint compound to the threads of the tailpiece; if you're using a metal waste, screw the tailpiece into place. From there on, it is just like hooking up a lavatory drain.

As I said earlier, you probably will not be making many of your own plumbing installations, but knowledge is power. Knowing what your plumber is doing helps you to supervise the job better. Being able to step in and work with a plumbing fixture in a pinch is also helpful. Leave your plumbing to professionals when you can. Check their work carefully, because a mistake made by a plumber can be devastating to a new bathroom. With this said, we are ready to move to the next chapter and learn a little about electrical work.

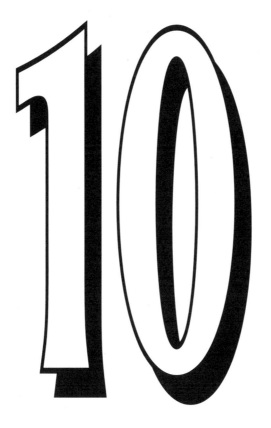

Electrical
considerations

ELECTRICAL considerations in bathroom remodeling are usually only a small part of the job. There are, of course, exceptions to this rule. Most bathroom remodeling electrical work involves little more than replacing light fixtures or adding a ground-fault-interceptor (GFI) circuit or outlets. Of course, if you are adding a new whirlpool, a new circuit will be needed to run it, and this must also be on a GFI circuit. The addition of a heat lamp, ventilation fans, or additional lighting fixtures can also run up the amount of electrical work to be done. Relocating existing electrical wires may also be needed if walls are being torn down, modified, or added. Dealing with electric baseboard or wall heaters can also add to the electrical burden. But generally speaking, electrical work in a bathroom doesn't usually amount to much.

Old bathrooms are usually pretty skimpy on electrical outlets. This can prompt your customer to have new outlets added. This is not a big job, but it is an issue you must address in the planning phase of the work. Code requirements may force you or your customer to add outlets, especially if you will be adding a new counter-top area in the room. The best bet when it comes to electrical work is to call in your electrical contractor to look the job over. Since electrical work is a trade where a license is required, like with plumbing, you will need the services of a licensed electrician if you are doing more than simple replacements.

I have wired entire homes that I've built for myself, and I've done plenty of replacement work on my own, but electrical work can be dangerous, and it is best left to licensed electricians. You can, however, do some phases of the work yourself, if you know what you are doing and can do it safely. For example, you could remove existing fixtures yourself and cap off the wires during demolition work. If all you are doing is replacing a light fixture or two, you can probably do that yourself. Once you get into adding new electrical devices, however, you will need a permit for the work and a licensed electrician.

Electrical services

Sometimes homeowners either want to or have to upgrade their electrical services for major remodeling projects. This is not normally the case in bathroom remodeling, but it can be. When a homeowner knows that an electrician will be working on a job, it is safe to assume that additional work, unrelated to the remodeling, could be done at the same time at a cost lower than what a special trip would be. This makes sense, and it is usually true. Replacing an electrical surface could be required in bathroom remodeling if new circuits are needed and the existing control panel is full. This, of course, is not a job you should do yourself, unless you are a well-trained, licensed electrician with experience in working with panel boxes. All work with electricity poses some danger, but the risks involved with replacing a service panel are too great to take.

It has been some time since I was involved in a job where the electrical service was to be upgraded, but the work was expensive then, and it must cost more now. If you are bidding a job, don't think that upgrading a panel is a cheap proposition that you can throw in to win a bid. Have your electrical contractor inspect the job and provide you with a firm quote before you give your customer a price.

Installing new electrical boxes

Installing new electrical boxes is not a complicated procedure. However, new installations usually require a permit and the work of a licensed electrician, unless homeowners are doing their own work. Poor workmanship with electrical wiring can result in fatal shocks and houses being burned to the ground. Some contractors allow homeowners to participate in the work involved with remodeling jobs. I've done this in the past with such phases of work as painting and found the risk to outweigh the advantage. When you let homeowners work on jobs where your name is associated with the work, you can have your reputation damaged when other people see the quality of the work. The decision of whether or not to allow the homeowner to participate in a job is up to you, but I advise against it.

Before electrical boxes can be installed, the type and size of box needed must be determined. The size and shape of electrical boxes vary with their purpose. To gain a better understanding of why there are so many options in electrical boxes, let's look at each type and see when it might be used.

Switch boxes

Switch boxes are usually rectangular. These same boxes are commonly used for wall outlets and wall-mounted lights. The dimensions for rectangular boxes are generally three inches by two inches. These boxes might be made of metal (Fig. 10-1) or plastic. Most jobs have plastic boxes installed. When this type of box is used, the ground wires must be mated under a wire nut. Metal boxes have green screws that ground wires can be attached to. From my experience, plastic boxes that come equipped with nails already implanted in them are the fastest type of box to use, and they are inexpensive.

Boxes for ceiling lights

Boxes used for ceiling lights are often octagonal. These boxes may also be used as junction boxes for joining numerous wires together. Each side of the boxes are typically four inches long. Round boxes are also used for ceiling lights. These boxes come in different varieties. For example, you can buy a box that will attach directly to a ceiling or floor joist. Another style has a sliding bar that can be attached to two joists and will allow the electrical box to be positioned anywhere between the joists. The advantage to the sliding boxes is the versatility with their placement.

Junction boxes

Both octagonal and square boxes are used as junction boxes. Square boxes are more common and have typical dimensions of about four inches. These boxes are used to join or extend wires where no

Figure 10-1

A metal electrical box. Photo by Images International, a division of Lone Wolf Enterprises, Ltd.

electrical device is installed. The boxes are covered with a solid plate to retain the wiring connection safely.

Depth requirements

Depth requirements for electrical boxes are determined by the number of wires to be placed in the box. Common depths vary from just over one inch to about three and one-half inches. It is usually best to buy boxes a little larger than what you need. This allows more space for bending and pushing wires into the box, which can make an installation easier.

Mounting electrical boxes

The mounting of electrical boxes can be done in a number of ways. Some boxes are sold with nails already inserted in them; all you have to do is position the box and drive the nail into a piece of wood. Other boxes have flanges that nails are driven through. Some boxes have flanges that move and allow more flexibility in where the box can be installed. Most electricians use plastic boxes with nails already in them for switches and outlets. Metal boxes are common for overhead electrical devices.

Boxes for ceiling fixtures are often nailed directly to ceiling joists. If the boxes need to be offset, such as in the middle of a joist bay, metal bars can be used to support the boxes. The metal bars are adjustable and mount between ceiling joists or studs. Once the bar is in place, the box can be mounted to the bar.

Rough-in dimensions

Rough-in dimensions can be determined by local code requirements and the fixtures to be served. There are, however, some common rough-in figures you may be interested in knowing about. Wall switches are usually mounted about four feet above the finished floor. Outlets are normally set between twelve and eighteen inches above the floor and are spaced so that there is not more than twelve feet between the outlets. When outlets are being installed along a counter, the distance between outlets should be reduced to no more than six feet.

Where does the red wire go?

Where does the red wire go? What should be done with the black wire? These are questions many people have about wiring. Electrical wires are insulated with different colors for a purpose. The colors indicate what the wire is used for and where it should be attached. Black wires and red wires are usually hot wires. White wires should be neutral, but they are sometimes used as hot wires. Don't trust any wire not to be hot. Green wires and plain copper wires are typically ground wires.

When matching colored wires to the screws in an electrical connection, they should be installed with a color-coding approach. Black wires should connect to brass screws. Red wires should connect to brass or chrome screws. White wires are normally connected to chrome screws. Green wires and plain copper wires should connect to green screws. Electrical wires should be crooked and placed under their respective screws in a way that the crook in the wire will tighten with the screw. In other words, the end of the crooked wire should be facing in a clockwise position under the screw.

Wire nuts should be used when wires are twisted together. The colors of wire nuts indicate their size. Refer to instructions on the box containing wire nuts to see what size wiring they are approved for. Wire nuts are plastic on the outside and have wire springs on the inside. When wires are inserted into the wire nut, the nut can be turned clockwise to secure the wires. It is important to use a wire nut of the proper size, and it should be installed to a point where no exposed wiring is visible.

Ground fault interrupters

Ground fault interrupters (GFIs) are generally required in locations where a source of water is close to an electrical device. GFIs are safety devices; they kill the power to an electrical device if moisture is detected. It is possible to install GFI outlets or GFI circuit breakers. Check with your local code enforcement office to determine the requirements in your area for GFIs. Be aware, though, that bathrooms do require GRI protection.

Ventilation fans

Bathrooms that do not contain windows that can be opened are required to have ventilation fans. These fans mount in the ceiling and may contain a light, in addition to the fan. The light can be a regular light for illumination, or it may be a heat lamp. Many homeowners don't like the idea of having a fan installed. Complaints range from appearance to noise when the fan is running. However, this is a code requirement that you must comply with. Styles and designs of exhaust fans vary widely (Figs. 10-2 and 10-3), so you should be able to find a unit that your customer will accept.

If you are installing new electrical fixtures in a bathroom, such as a fan or whirlpool motor, your electrician will probably have to run wires from the bathroom to the service panel. This is something that you should plan for in the bidding stage of your jobs. Sometimes the routes required for wiring are easy to maneuver, but there are times when it is very difficult to get new wires to an existing panel box. Prepare your customers for this in advance. Don't wait until the job is under way and your electrician tells you that there is a problem that will involve more demolition than what you and your customers had planned for. Find out before the job starts what the electrician will need to do, and explain the needs to your customers.

Recessed lighting

Recessed lighting can be very attractive in a bathroom, especially a large bathroom where a lot of artificial light is needed. Lights and outlets are required to be on different circuits. It's possible that your electrician can tie into an existing circuit when adding new fixtures and devices, but you cannot be sure until you consult your electrician. Do this early on in the bidding phase. Recessed lighting (Fig. 10-4) can be very nice over large whirlpool tubs, in alcoves that contain toilets, and over vanity areas. Not many contractors think about alternative or creative lighting in a bathroom, and this is a design mistake. Bathrooms are one of the most-used rooms in a home, so they should be both attractive and functional.

Track lighting

Track lighting is not often associated with bathrooms, but there is no reason why it shouldn't be. The advantages of track lighting are many. For example, the lights can be positioned to produce light in various sections of the room. A gang of track lighting (Fig. 10-5) can produce

Figure 10-2

Deluxe Fan-Light Combinations

8663M Satin Aluminum
8663MAB Antique Brass
8663MBR Polished Brass

8663F
Fluorescent

8663DG
Round Genuine Wood

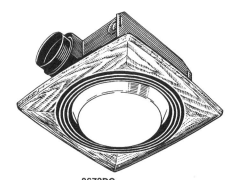

8673DG
Square Genuine Wood

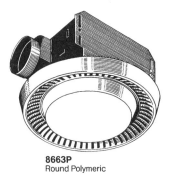

8663P
Round Polymeric

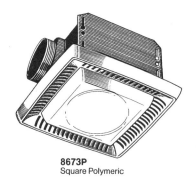

8673P
Square Polymeric

Deluxe fan-light combinations. Courtesy of Nutone

Figure 10-3

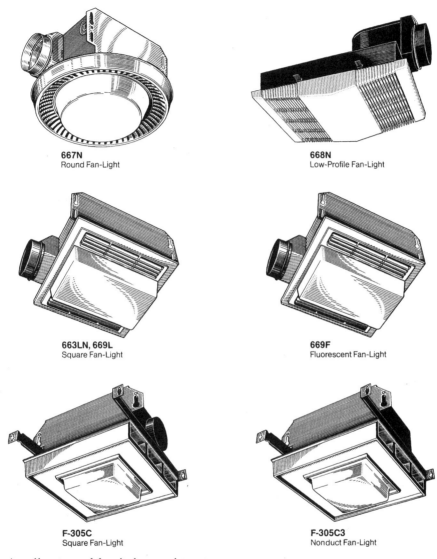

667N
Round Fan-Light

668N
Low-Profile Fan-Light

663LN, 669L
Square Fan-Light

669F
Fluorescent Fan-Light

F-305C
Square Fan-Light

F-305C3
Nonduct Fan-Light

A collection of fan-light combinations. Courtesy of Nutone

Figure 10-4

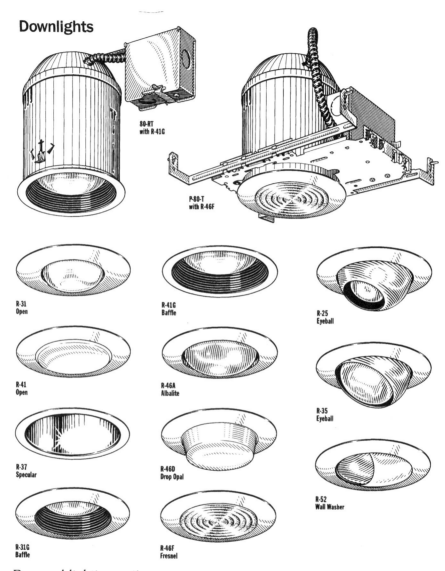

Downlights

80-RT
with R-41G

P-80-T
with R-46F

R-31
Open

R-41G
Baffle

R-25
Eyeball

R-41
Open

R-46A
Albalite

R-35
Eyeball

R-37
Specular

R-46D
Drop Opal

R-52
Wall Washer

R-31G
Baffle

R-46F
Fresnel

Recessed lighting options. Courtesy of Nutone

Figure 10-5

Low-Profile Track

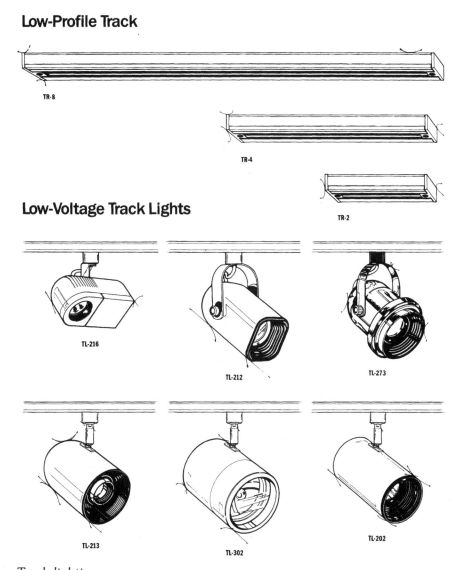

TR-8

TR-4

TR-2

Low-Voltage Track Lights

TL-216

TL-212

TL-273

TL-213

TL-302

TL-202

Track lighting. Courtesy of Nutone

a lot of light, which is especially pleasing in a make-up area. Installation of track lighting is easy, and the lighting fixtures are not too expensive. In addition to their functional abilities, track lights make a decorative statement in a bathroom.

Wall-mount fixtures

Wall-mount fixtures have been common throughout the years in bathrooms. It is not unusual to find a wall-mounted light on either side of a mirror or medicine cabinet. The days of having a fluorescent light tube over or on each side of a medicine cabinet are all but gone. Wall-mount lights are available in many styles that allow a designer to match the lights to the theme of a bathroom. As a remodeler or builder, you should try to think in terms of themes. Don't think of bathrooms as simple spaces where necessary duties are performed. Consider the rooms retreats. These retreats should be tasteful, relaxing, and inviting. You can accomplish part of this goal by choosing proper lighting fixtures.

Strip lights

Strip lights are probably the most popular form of lighting fixture installed in modern bathrooms. The strips may have a wood, chrome, or brass finish (Figs. 10-6 and 10-7). Bulbs used in these strips are usually large. They are sometimes clear and sometimes frosted. Strip lights produce a lot of heat, and this is one drawback to them, but they also produce a lot of light. You might install strip lights above and on both sides of a vanity mirror. This is both functional and decorative. Strip lights are very affordable, and depending on the finish and the bulbs used, they can be very complementary of the bathroom.

Recessed heat lamps

Recessed heat lamps are quite popular with many people. I've never installed them in any of my homes, but I have used them in motels. It is nice to have the warmth of the light as you towel off after a bath or

Figure 10-6

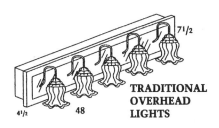

TRADITIONAL OVERHEAD LIGHTS

7¹/₂   4¹/₂   48

TL48 (5 lights)

Cut out dimensions for
TL48 are 45¼ x 4¾.
Elite Series is white painted hardwood.

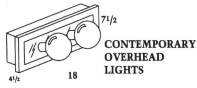

CONTEMPORARY OVERHEAD LIGHTS

7¹/₂   4¹/₂   18

CL18 (2 lights)

Cut out dimensions for
CL18 are 15¼ x 4¾.
Elite Series is white painted hardwood.

**Bulbs for Contemporary Lights are not included.
All Overhead lights will face or recess mount.**

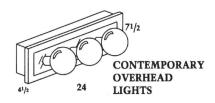

CONTEMPORARY OVERHEAD LIGHTS

7¹/₂   4¹/₂   24

CL24 (3 lights)

Cut out dimensions for
CL24 are 21¼ x 4¾.
Elite Series is white painted hardwood.

CONTEMPORARY SIDE LIGHTS

7¹/₂   4¹/₂   8

C2423

Shipped in pairs.
Side light will face
mount only.
Elite Series is white painted hardwood.

Vanity lights.   Courtesy of Nutone

shower. Heat lamps (Fig. 10-8), however, require a lot of power, and this can strain a crowded circuit. Talk with your electrician before you suggest heat lamps to your customers. The addition of heat lamps might be all it will take to throw you into a need for a new circuit that could otherwise be avoided. If a heat lamp is to be installed, you should consider a combination heat lamp and fan, as I described earlier.

Electric heat

Electric heat is sometimes used in bathrooms as either baseboard heat or as a wall heater. If you are installing this type of heat where it has

Figure 10-7

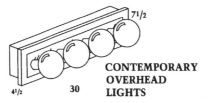

CONTEMPORARY
OVERHEAD
LIGHTS

CL30 (4 lights)

Cut out dimensions for
CL30 are 27¼ x 4¾.
Elite Series is white painted hardwood.

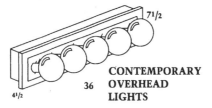

CONTEMPORARY
OVERHEAD
LIGHTS

Strip lights. Courtesy of Nutone

CL36 (5 lights)

Cut out dimensions for
CL36 are 33¼ x 4¾.
Elite Series is white painted hardwood.

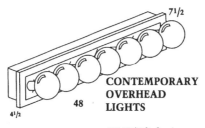

CONTEMPORARY
OVERHEAD
LIGHTS

CL48 (7 lights)

never been before, you can count on running a new circuit. Electric wall heaters (Figs. 10-9 and 10-10), with blowers, are quite common in bathrooms. These units mount recessed into a wall. If this will be a new installation, you have to make sure that the wall cavity will accept the heater, and you will need new wiring run for it. Replacing existing electric heat or heaters should not require new circuits, but you must confirm that the existing wiring is adequate for the newer heat. Pay

Figure 10-8

QT-9093 Series Quiet Test
Heat-A-Ventlite®

HST-24
Programmable
Switch

9965 Heat-A-Ventlite®

9093 Series Deluxe
Heat-A-Ventlite®

9960 Heat-A-Lite®

9013NL Heat-A-Lite®

9905 Heat-A-Vent®

Heat lamps and heat-lamp-fan combinations. Courtesy of Nutone

Figure 10-9

Wood-grain electric wall heater. Courtesy of Nutone

Figure 10-10

Standard electric wall heater. Courtesy of Nutone.

close attention when you are estimating jobs that involve electric heat and always have your electrician assess the existing conditions before you commit to a firm price with your customer.

Counter outlets

Counter outlets are often sparse in older bathrooms. Many old bathrooms have only the outlet that is built into a medicine cabinet. This is a part of the electrical layout that you can, and should, update. As technology rolls on, we are faced with more and more electrical appliances intended for use in bathrooms. Let's think for a moment of what some of them are.

A hair dryer may use around 1,000 watts of electricity. An electric toothbrush doesn't pull a lot of power, but it is one more thing to plug in. Oral irrigation devices are common, and they must be plugged in. Curling irons require an outlet. Electric hair curlers are another device that requires an outlet. Electric shavers have to be plugged in somewhere. Maybe an ultrasonic jewelry cleaner will find its way to the bathroom counter to be plugged in. We could go on and on with a long list of potential appliances and devices that require connection to outlets. With this being the case, the more outlets that are available, the more convenient the bathroom makeup and work space will be.

Code requirements for outlets in bathrooms are set as minimum standards. This doesn't mean that you can't add more than the minimal number of outlets required. If you are building or remodeling a bathroom where extensive counter space is available, you should consider increasing the number of electrical outlets in the counter area. Remember, the outlets have to be on a GFI circuit or contain GFI outlets. By installing additional outlets, you increase the overall cost of the job a little, but you make the bathroom a much nicer place to be.

The electrical aspects of bathroom remodeling are pretty minimal. A few outlets, a few lights, maybe a fan, and you're done. Well, it's not always quite that easy, but there really isn't a lot involved with electrical work in most bathrooms. The same principal usually applies to heating considerations in bathrooms. To illustrate this point, let's turn to the next chapter.

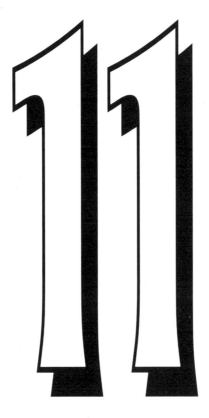

HVAC aspects of
a job

THE HEATING ventilation and air-conditioning (HVAC) aspects of a typical bathroom job are usually minimal. Bathrooms can have a number of different types of heat. Some bathrooms have radiators. Others have hot-water baseboard heat or electric baseboard heat. Electric wall heaters are often found in bathrooms, and yes, many bathrooms get their heat through HVAC systems. Your work with an HVAC system in remodeling a bathroom may not amount to anymore than removing floor grates and replacing them when you are done. However, you might find it necessary to add new registers in the room, especially if you are enlarging it. If you are building a bathroom addition, you may have to extend ductwork to the new living space. You may also find it necessary to relocate existing registers or grilles to accommodate new fixture locations. Usually none of this work is complicated.

Plumbing is normally responsible for most of the work in bathroom remodeling. There is a degree of carpentry work involved, and walls and flooring can account for some substantial work. Electrical work and HVAC work is usually the least of your worries. However, you can run into some circumstances where an HVAC will give you a little trouble. Fortunately, these cases are few and far between.

When a bathroom is served by forced hot-air heat, the heat is usually delivered through floor registers. It might enter the room through wall registers. Typically, the register grilles are the only part of the system that has to be dealt with. This is simply a matter of lifting the grilles out of the duct boot and protecting the boot so that foreign matter doesn't fall into it. As simple as this is, some contractors fail to seal off the ductwork and create problems for themselves. Since this does happen with some regularity, let's talk briefly about it.

A majority of heat registers are found in the floor. When you remove the grille the boot of the duct is left open. The opening is large enough for all sorts of items to fall into it. Some of the common debris includes nails, pieces of wood and drywall, trash, and so forth. When these objects get into the ductwork, they can create obstructions to the air flow. They can also cause unwanted noise when the system is in operation. Removing the debris might be as simple as reaching into the duct, or you may have to gain access to the supply line and disassembly it. No homeowner is going to appreciate a contractor tearing out a section of ceiling below a bathroom to remove objects that should never have been allowed access to the duct in the first place. Avoid this by sealing the duct opening.

You can block the opening of a boot in many ways. A simple, but not always effective way, it to apply duct tape over the opening. This will work until the tape is stepped on or punctured to a point where items can fall through it. Some contractors stuff the duct with fiberglass insulation. This works, but cutting a plywood insert to fit over the opening is probably the most secure way to block the duct. When you put plywood over the opening, you can be fairly sure that nothing is going to get into the ductwork. However you do it, make sure that you secure the duct to keep unwanted objects from finding their way into the duct.

Heating system modifications are not needed in most bathroom remodeling jobs. Unless you are expanding the size of the room, the existing heat should be adequate and require no major work. However, there are times when heat needs to be relocated within the room. Assuming there is good access from under the floor, moving heat around is not normally a big job.

Ducts used in forced-hot-air heating and central air conditioning has very sharp edges when disassembled. If you are going to work with heating or air-conditioning ducts, wear gloves. The ducts are usually held together by metal strips that slide into a channel and hold the ducts together. These strips can be dislodged with a hammer. Metal fabrication shops will be glad to make lengths of ductwork or offsets to your specifications. Other than for the risk of cutting yourself on sharp metal, installing duct work is not too difficult.

In some cases flexible ducts can be used to carry heat or cool air from a main trunk to an outlet register. Flexible duct is obviously easier to work with, and you are not as likely to hurt yourself. If you will be adding new ducts to an existing system or extending the length of existing ducts, talk with some professionals beforehand. You or your workers can probably do the job, but it helps to consult your HVAC contractor if you have any questions. Most subcontractors are willing to offer a little free advice on small jobs if they work regularly for you on larger jobs.

Typically, the size of duct work gets smaller as it goes along its route. If the duct is not sized properly, it cannot perform to its optimum output. These types of alterations may affect the effectiveness of the heating and cooling system. New ducts can be cut into existing trunk lines easily, but you must be sure your alterations will not strain the system.

One of the first considerations you must assess is whether you and your crews will be doing the work or whether you will engage an HVAC contractor. Obviously, if all you are doing is removing a few grilles, you shouldn't need the help of a subcontractor. Even making minor modifications can probably be done with your in-house workers. If you will be expanding a room or adding a new room to a home, you should at least talk with an HVAC expert for advice of sizing and routing the ducts.

Main trunk lines for ductwork are normally made of metal. But, many of the supply lines can be made of flexible duct. This is an easier material to work with, and it doesn't involve as many sharp edges to be wary of. Access is probably one of your major considerations if you will be relocating or adding ductwork. If you will be removing the subflooring during your remodeling, you should have adequate access to reroute existing ducts. When you can get under the room that is being remodeled, you often have excellent access. In any event, you have to plan in advance for how you will gain enough access to get the job done.

Extending new ducts to an addition can require you to penetrate a foundation wall. This doesn't have to be a complicated procedure, but it does add an element of work to a job that is not always present. You may have to work your way through a stone, block, brick, or concrete foundation. But, you may also be able to work the duct through the band board of the home to avoid this. Planning is the key. When you look over a job for an estimate, take the time to draw a routing map for all of your duct needs.

▲

Simple modifications

Simple modifications are just that: simple. Let's say that you are installing a new vanity to replace a wall-hung lavatory. The existing heat register is located under the existing lavatory. This installation has worked fine, since air can circulate under the open lavatory. But, when you cover the grille with a vanity, there will be no air circulation. This register needs to be relocated. There are two probable solutions to this problem. The most logical solution is to move the duct so that the register will be mounted in the floor to one side or the other of the new vanity. If for some reason this line of action was not feasible, you might extend the duct up, into the base of the vanity and allow it to blow air out one side of the cabinet, through a register.

Most customers will not want a majority of their vanity space taken up by an ugly piece of ductwork. This is motivation to find a way to move the boot to another floor location. When you are standing in the bathroom estimating the job, moving the register might seem simple. It probably will not be very complicated, but it will not be as easy as you might think. The supply duct will be running in the joist bay. To move the duct to one side or the other, you may have to cut a joist or two. This, or course, will mean installing headers in the cut joists to maintain structural integrity. Unlike a water pipe or an electrical wire that can penetrate a joist by just drilling a hole, a duct is much larger and requires the removal of more wood. Yet, the job can be done. If you have a crawlspace or basement under the bathroom, you might be able to relocate the duct without excessive cutting, so check your options carefully.

Assuming that you are going to move the boot for the register, you will need some fittings. HVAC suppliers keep a stock of standard fittings available. If you need a unique angle, any good metal shop can make a fitting for you. Converting from metal duct to flexible duct is another potential option. One way or another, you are going to have to relocate the boot. The work will not be complicated or unusually difficult, but it may be time-consuming, so allow for it in your bid.

If you have no spare floor space to work with and conclude that the duct must come up in the vanity, you can simply extend a vertical section of duct from the existing duct and then connect it to a wall register that will be cut into the side of the vanity. I would avoid this approach if at all possible. Putting the duct in the vanity will work just fine, but it does destroy valuable storage space and a new vanity with a heat duct in the side of it is not a thing of beauty.

Is there another option when floor space is at a premium? Consider extending the existing duct to a point where it can turn up in the bathroom wall and provide its air circulation through a wall register. This will eliminate the clutter in the vanity, and the register will look more at home on the wall than it would on the cabinet. You should be seeing already that there are a number of options open to you when working with ducts.

Adding new registers

Adding new registers during the remodeling of a bathroom is unusual. Unless you are expanding the size of the room considerably, there should be no need for this type of work. However, since you might run into a situation where this would be necessary, let's talk a little about the basic procedures involved with the process. Let's assume that your customer wants a much larger bathroom. For our example, we will say that a walk-in closet is going to be converted to bathroom space. Due to the existing layout, all you will have to do is knock out a section of interior partition wall that separates the closet from the bathroom. This will give you an L-shaped bathroom space to work with. There is, however, a problem. The walk-in closet was never equipped with supply ducts for heating or air conditioning. This was not a problem when the closet was a closet, but now the space is going to used as a part of the bathroom. The existing duct register in the old bathroom is not going to be adequate to heat and cool the added space. What are you going to do?

Your options in this case are somewhat limited. It's possible that you could branch off an existing supply duct from either the bathroom or the bedroom to install a new register in the floor area of the converted space. The problem with this approach is that the ducts were sized for the job that they are doing. They were not meant to be large enough to carry adequate air flow for two registers. Tapping into the existing supplies will provide some air circulation, but probably not enough. Also, the tap-in will cut down on the amount of air reaching the register served by the duct originally. What you should do is run a new supply duct from either the trunkline duct or from the plenum. This can get tricky, due to access.

Supply ducts originate from either the plenum, which is near the furnace, air conditioner, or heat pump, or from a main section of duct that is called a trunkline. The heating or cooling source can be located almost anywhere in a home. Trunklines extend outwardly from the heating or cooling source and get smaller as they go along. In homes with basements or crawl spaces, the trunkline is easy to see and identify. There are times, however, when the trunkline is concealed. This makes finding it and attaching to it difficult. Even once you find a suitable source to connect a supply duct to in a remodeling situation, routing the duct to the place where a register is wanted can be very difficult. Electricians can snake wires effectively, and plumbers can work some magic with small pipes, but supply ducts are too large to hide easily.

If access is not a problem, connecting a new supply duct to an existing trunkline or plenum is easy. But, when there is no obvious route to take the new duct, a change in plans might be in order. For example, you might consider adding an electric wall heater in the new part of the bathroom. This will require a new electrical circuit, but it should be much easier to run than a new duct if you don't have open access for the duct installation. Customers may be willing to have you cut out a portion of a ceiling to gain the access you need to install ducts. If this is the case, the job can be done with relative ease. You will have to talk with your customers to see how much disruption they are willing to go through to get new heating supplies in the converted space.

Heating systems are sometimes mounted in attics. This is not common, but it is done. If you are working with a remodeling job on the top floor of a house, an attic-mounted system makes access to the trunkline easy. If you haven't found the heating system in or under the home, look in the attic, it could be there.

Additions

Additions to homes can create more living space than an existing heating or cooling system can handle. If you are going to be building a large addition, say for a master bedroom and a master bathroom, you should have an HVAC contractor evaluate the existing system to see if it is large enough to take on the added living space. If it is, you can have new supply ducts run from the existing system. Access and routing are again the only problem with this type of a job, assuming that the existing system is large enough to take care of the new space.

What if the new addition is too large for the existing furnace, air conditioner, or heat pump? Are there any options beside electric heat or replacing the main system in the home? Yes, you could install a stand-alone heat pump for the new addition. This won't be a cheap proposition, but it will be effective and less expensive than a major retro-fit on the main system. Heat pumps are available in a variety of sizes. Some of them mount in a wall and are intended to meet the needs of a single room. You may have seen this type of heat pump in motels or hotels, they are common in such places. A two-piece heat pump has one piece of equipment outside of the building and another piece inside. This is the type of heat pump normally used to heat and cool an entire home. Either type might be appropriate for your needs in the addition. Talk with your HVAC contractor for detailed information on heat pumps and your choices in working with them.

Simple ventilation

The most simple aspect of ventilation in a bathroom can be the installation of a bathroom fan that exhausts to the exterior of the home (Fig. 11-1). There are combination units that offer some heating along with ventilation (Fig. 11-2). Electricians are often the contractors who install bath fans. However, your HVAC contractor may consider the fan a part of ventilation, and therefore, a part of the HVAC system. You should clarify with your subcontractors who will be providing bath fans. You can choose between straight ventilation units, heat-and-vent units, light-fan combinations, and concealed-

Figure 11-1

671, 672
Ceiling Fans

686
Vertical Discharge Fan

696N, 693, 695,
C-350C2 Bath Fans

C-350A Bath Fan

NUA-40, C-305C3N
Non-Duct Fans

Bathroom ventilation fans. Courtesy of Nutone

Figure 11-2

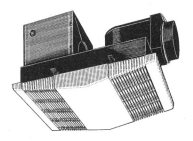

665N Heat-A-Ventlite®
665S Heat-A-Ventlite® includes switch

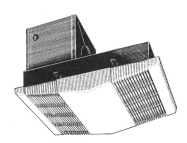

Heat-fan combination units. Courtesy of Nutone

660N Heat-A-Lite®

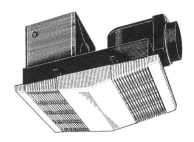

605N Heat-A-Vent®

intake fans (Fig. 11-3). My guess is that your electrician will probably install these units, but you should make sure who is doing what before you accept a final bid from your subcontractors.

The odds of running into major HVAC work in bathroom remodeling are low. You may remodel dozens of bathrooms without

Figure 11-3

Concealed intake bath ventilation fan. Courtesy of Nutone.

8833
Concealed-Intake Bath Fan

ever encountering more than removing register grilles. When you find yourself in a more complicated situation, call in your HVAC subcontractor. If the work gets to a point beyond what we have discussed here it will most likely be out of your level of expertise. Don't take chances on ruining an otherwise good job by trying to save a few dollars in cutting out the heating or cooling contractor. Work within your known limits, and call in outside professionals when you sense the work is beyond your capabilities.

Flooring options

THE FLOOR of a bathroom is a focal point of the room. The right floor coverings can distinguish the room and make a statement about its owner. The wrong floor covering can darken the room or give an unbalanced appearance. The choice in a finished floor covering will have a strong affect on the overall appearance of any remodeled room. Bathrooms tend to be small, so people looking at the room have more opportunity to take in all the nuances. Flooring, walls, plumbing fixtures, and light fixtures are what will most often be noticed. Selecting the right floor is important for any room, but it is critical for a bathroom. The type of flooring used and the pattern created can have a major impact on how a bathroom will look. There are certain aspects of bathroom remodeling that deserve more attention than others, and flooring is certainly one of them.

Even with the best finished floor covering available, bad subflooring and floor joists can prohibit a remodeled bathroom from being outstanding. If the floor squeaks every time you walk across it, you will notice the squeak more than you will the attractive floor covering. If the joists are weak and the floor is spongy, you may wonder when your bathtub is going to fall through it. Cutting corners on the flooring in a bathroom is a mistake.

Most of the work done in routine bathroom remodeling is not of a structural nature, but the flooring is. While you don't need superior framing skills to build a linen closet, you may need them to repair damaged floor joists. It is not all that uncommon to discover rotted or damaged floor joists when remodeling a bathroom. Little leaks can do a lot of damage over time. A toilet that condensates or leaks around its flange can ruin a floor structure fairly quickly. A faucet for a bathtub or shower can leak slowly in a wall for a long time before the leak is discovered. During this time, moisture damage is occurring. What would you do if you removed water-stained subflooring and found the tops of three floor joists rotted to a point where the point of a pencil would penetrate them? A lot of people would panic and assume the joist would have to be removed and replaced. It is possible that the joists would need replacement, but it is more likely a little repair job could solve the problem. Homeowners often panic when such damage is discovered. Professional contractors should know enough not to be too concerned. Yes, the problem will have to be

▲

219

corrected, but the repair procedures are not particularly difficult or expensive. Be prepared to ease the minds of your customers if major structural damage is discovered in subflooring or the joist system.

Many readers of this book are accomplished carpenters, but just as many may be business owners who don't have a full knowledge of the carpentry and flooring trades. For this reason, we will address issues with an assumption that readers are not experts in this phase of remodeling. You will learn about floor joists, subflooring, underlayment, and finished floor coverings. There will be tips on how to handle rotten joists the easy way, and how to get the bubbles out of new vinyl flooring. You may know some or even much of what we will discuss, but there will probably be some tricks and ideas that you have not yet learned.

Floor joists

Floor joists are the structural members that support the subfloor. They are normally boards with dimensions ranging from 2" × 8" to 2" × 12". Floor joists span the distances between outside walls and support girders. The length of the span and the use of the floor influences the size of the joist. When floor joists are sized, two types of weight loads are taken into consideration. There is a dead-weight load and a live-weight load. Dead-weight is considered to be objects such as furniture or fixtures. Live-weight is the weight of people occupying the room.

If you remove subflooring and find a few of the floor joists to be rotted, you may not have to replace them. It is entirely possible that you can add new supports, without removing the old ones. In many cases all you will have to do is slide new joists in place on each side of the damaged joist and nail them to the old joist. The new joists should be the same length and have the same dimensions as the old joist. This is normally not a very difficult procedure and it works fine in most circumstances. However, the banging on the joists and nails can create problems in a finished ceiling below the room.

If a ceiling is attached to the joists that you are working with, it is very likely that the vibrations from your work will cause drywall nails or screws to work loose and pop out the joint compound that was hiding them. This means that you will have to patch and paint the ceiling below. A situation like this is difficult to predict when you are estimating a job. Unless you can see obvious water damage, and sometimes you can, you have to assume that the joists and subflooring will be solid. One way to protect yourself from having to do extra work at your quoted price is to address the issue in your proposal and contract to the homeowner.

Some contractors don't like to raise the question of potentially bad circumstances when selling a job. If you paint too horrible of a picture, you may scare the customer off. This is true, but leaving too many possibilities unquestioned can result in lost money for you. This is a big enough issue that we should discuss it a little further.

If you will be remodeling a bathroom that has either no ceiling or an unfinished ceiling below it, you can skip exclusions about nail-pops and repainting ceilings below. You are still left with the possibility of having to supply and install replacement joists and subflooring. If you have bought building materials lately, you know how much lumber and plywood have increased in price over the years. Would you rather gamble on paying for these materials out of your profits or addressing the possibility with your customers? Personally, I would bring the issue up in my discussions with the customers. I would explain that there is no apparent evidence to indicate additional cost, but that if damage is found during demolition the repairs will be charged out as an extra, over and above the contract amount. This approach has worked successfully for me over the years. The choice is yours, but if you don't bring it up before going to contract, you may have to pay out some of your profits for unexpected work and materials.

If only a small section of a joist is damaged, you may be able to get by with scabbing new pieces of wood on the old joist. Assume an existing joist was damaged, but only for about two feet of its length. It may be that you could attach new wood on each side of the damaged area to avoid installing complete joists. The scab wood should extend well past the damaged area; in this case, scab wood about four-feet long should be sufficient. This practice may not be suitable in all situations; check with your local building inspector before relying on this type of repair.

Another option available for joists with small areas of damage is a process where the old joist is headed off. To head off a joist, you cut out the damaged section. Then joist-size material is used to span the distance between sound joists. The cut end of the damaged joist is attached to the new wood that is running perpendicular to the cut end. This procedure is also used when an opening is needed between floor joists for the passage of chimneys or stairs.

Subflooring

Subflooring is the flooring that attaches to floor joists. It can be made of boards, but it is usually made with sheets of plywood, waferboard, or particleboard. Most jurisdictions allow two options when installing subflooring. One layer of $\frac{3}{4}$" tongue-and-groove material is allowed or two layers of standard $\frac{5}{8}$" plywood, waferboard, or particleboard may be used. It is not normally acceptable to install a single layer of material that is not fitted with a tongue-and-groove installation. Many contractors frown on using waferboard in a bathroom floor structure. The concern is that the waferboard may become wet and delaminate. This will not happen if the bathroom floor is impervious to water and sealed properly.

Underlayment

Underlayment is usually a thin (about $\frac{1}{4}$") sheet of plywood that is laid over a subfloor. The underlayment is normally sanded on one side and provides a smooth surface for the installation of finish flooring. Underlayment is not always needed in a remodeling job. The existing floor may be in good shape and not require a new covering. However, when I figure a job that will be using vinyl flooring, I also calculate in the cost of new underlayment and crack filling. The surface required for an excellent installation of vinyl flooring has to be nearly perfect, and few existing floors meet this requirement.

Vinyl flooring

Vinyl flooring (Fig. 12-1) is the most common type of flooring used in bathrooms. It is generally available in widths of six or twelve feet. Vinyl flooring can be tricky to install, but the job can be done by anyone with average skills and patience. I always sub my vinyl work out to a professional, but I've known many contractors who install the flooring themselves. My installer gives me good prices on both the material and the labor, so I'm not inclined to mess with having my crews do the work and possibly ruining the material and my relationship with the customer. By having a subcontractor do the work, I have someone to turn to if the floor bubbles or is otherwise unsatisfactory. If you plan on doing the installation yourself (Fig. 12-2), make sure that you know what you are doing. One wrong cut can basically ruin the entire floor and force you to buy all of the material

Figure 12-1

Samples of sheet vinyl flooring. Courtesy Armstrong World Industries, Inc.

Figure 12-2

Plan cuts on vinyl carefully. Courtesy Armstrong World Industries, Inc.

for a second time. If this happens, any money that you had hoped to save is down the drain, so to speak.

Before you start installing a new vinyl floor, make sure that the area of the installation is clean and smooth. The surface should be flat and without cracks, depressions, or bulges. Cracks in floors can be filled with special compounds. These filling compounds are available from the same stores that sell flooring. Don't attempt to cut a corner by not filling the cracks. Even tiny cracks can show through vinyl flooring. I think it costs me about $75 per floor to have my installer fill all cracks to perfection. This is money that I might be able to save, but the risk is not worth it to me. If the floor is installed and the cracks left under the flooring show through, the only option is to remove the flooring and start over.

Before installing your floor, roll the flooring up with the finish side facing outward. Leave it in this position for a full day. Maintain an even temperature of about sixty-five degrees F. in the room where the vinyl is being stored. This preparation work is important. Don't skip this step or the flooring may not lay well on the underlayment.

If the floor is going to require seams, make them before installing the flooring. Lay two pieces of flooring in place so that they overlap. Make sure the pattern meets and matches. Using a straightedge and a utility knife, cut through both pieces of flooring where the seam will be made. Remove the scrap flooring and attach the two pieces of finish flooring to the floor at the seam. Use a hand roller to press the flooring down. The back of the flooring should be in contact with the adhesive or tape you are using for the installation. Cover the seam with a sealing compound. With a few exceptions, you should not run into a bathroom where a floor has to be seamed. Avoid seams whenever possible, since they are a potential trouble spot down the road.

Vinyl flooring should be laid out in a room with enough excess vinyl that the flooring rolls up on the walls. A utility knife is one of the best tools to cut vinyl flooring with (Fig. 12-3). It is best to install the vinyl before vanities and toilets are set in place. You will have to cut the flooring around the bathtub or shower, but making as few cuts as possible will add to your chances of success. Even if you are very proficient at cutting around fixtures, the flooring is less likely to roll up if the vanity and toilet are sitting on it, rather than having the floor butt up to the fixtures.

When you are securing your flooring to the subfloor, you may use adhesive, tape, staples, or a combination of them all. A floor roller should be used to roll wrinkles out of a new installation. Rollers can be rented at tool rental centers. When the vinyl is flat, cut away excess flooring. Run a utility knife along a straightedge to cut the vinyl where it meets the walls. Baseboard or shoe molding (Fig. 12-4) will then be installed to hide the joint between the floor and wall.

Some builders and remodelers install vinyl flooring before they install their baseboard trim. In doing this, you eliminate the need for shoe molding, which saves some money. The downside to installing baseboard after the flooring is in could be that a painter will get paint or stain on the new floor. My preference has been to install baseboard

Figure 12-3

Cutting vinyl flooring is easy with a utility knife. Courtesy
Armstrong World Industries, Inc.

Figure 12-4

Fig. 10

*Shoe molding is installed in front of
baseboard molding to finish off the
installation of vinyl flooring.* Courtesy of
Georgia Pacific

before vinyl and then to install shoe molding to hold the vinyl down. I've seen some contractors who just allow the vinyl to butt up against baseboards, without using shoe molding. In a perfect world this will work, but it is a risk. First of all, cuts must be absolutely perfect for the seams at baseboards to look acceptable. Another risk is that the vinyl may pull loose and curl if molding is not holding it down.

Choosing vinyl flooring

Choosing vinyl flooring can be a long, tedious job. First you must decide on individual square tiles (Fig. 12-5) or rolled sheet goods. Most customers and contractors opt for sheet goods, as I do. Even with this decision made, choosing the right colors and patterns can take hours. Looking at the small sample squares offered by installers

Figure 12-5

Vinyl flooring is available in squares. Courtesy Armstrong World Industries, Inc.

and suppliers can be deceiving. The patterns of the samples are small, and this can lead to misconceptions of what a full floor will look like. Also, the light under which the flooring samples are observed can be very different from what the lighting will be in the bathroom where the flooring is to be installed. These two factors can combine to make for disappointed customers.

Is there some way around the problem with samples? There are no easy solutions. One method that I have used works, but it can take time to develop. When I do a job and there is scrap flooring left over, I save sections of it that are larger than regular samples. My installer saves pieces for me from other jobs. After a while of doing this, I have amassed quite a collection of larger samples. If you want to try this without investing the time in it that I have, you should contact local suppliers in your area and see if they would be willing to sell or give you small cuts from their products. The larger the sample is, the easier it is for your customer to envision what it will look like once it is installed (Fig. 12-6).

The cost of vinyl flooring covers a large range. Vinyl flooring material is priced by the square yard, and low-end materials can cost less than half of high-end materials. Helping your customers wade through this swamp of indecision can also be very time-consuming. You can take the approach that many contractors do and just send your customers to a showroom to pick their own materials. A lot of contractors operate this way, but I feel a personal involvement with the customers is helpful during the course of a job. Maybe I invest too much time in my customers, but without good customers, few contractors are going to survive in business.

How can you help your customers decide on a flooring? I don't think you should make definitive statements on which type of flooring they should buy. Just being with your customers during the decision-making stage is enough to show that you are taking a personal interest in them and their job. Low-end vinyl can look about as good as high-end vinyl when it is installed. Options that may influence your customers are as follows:

➤ Colors

➤ Patterns

➤ Durability

➤ Cushioning

➤ Shine

➤ No-wax finishes

➤ Resistance to cuts

➤ Depth of patterns

In the end, the salesperson at the showroom will have a lot of influence on many customers. Your presence may slow the

Vinyl makes a beautiful floor covering, but choose from the largest samples available. Courtesy Armstrong World Industries, Inc.

salesperson down from fluffing a product too much, and this is helpful for your customer. You can offer advice on your past experience with specific types of flooring. If the showroom has large rolls of flooring on display, you and the salesperson can roll the flooring out to give customers a good look at what their finished floor will look like. Be prepared to spend some serious times with your customers on some selection issues, and consider the flooring one of those issues.

Carpet

Carpet is not a common floor covering for a bathroom, but it is sometimes used. If your customer wants you to install carpeting in a bathroom, weigh the pros and cons with the customer. While carpeting makes the floor warmer and less slippery than vinyl or tile, it retains moisture, and that can create problems with mold, mildew, and rot. Some code officers frown on the use of carpet in a bathroom, since code usually requires a floor that is impervious to water.

If after careful consideration you decide to install carpeting in a bathroom, make sure the pile of the carpet faces the entrance; this will enhance the appearance of your room. Most carpet is available in widths up to twelve feet, so you should not need a seam in a bathroom.

The installation of loop-pile and cut-pile carpeting can be done by anyone with average skills, but I suggest leaving the job to professionals. If you or your crew will be installing carpet you will probably have to rent a few tools. You will also have to be careful in your measurements and cuts.

Most carpet is held in place by tackless strips. These strips are normally about four feet long and have sharp teeth that bite into the carpet. Tackless strips come in different sizes, check with your carpet supplier for the proper size to use with the carpet and pad you are installing. The tackless strips are installed around the perimeter of the area to be carpeted. Doorways and cased openings are sometimes fitted with metal trim strips. These strips are either folded over or nailed on top of the carpet to give a finished edge that people will not trip over.

Tackless strips should be installed with a uniform distance from the wall. Check with your carpet supplier for the proper distance to maintain between the edge of the strip and the wall. A rule-of-thumb distance is a gap equal to two-thirds the thickness of the carpet. The carpet pad is installed within the boundaries of the tackless strips, but it is not attached to the strips. Check the manufacturer's recommendations to determine which side of the pad should face the subfloor. Carpet pads are usually stapled to the subfloor.

Carpeting should be unrolled, flattened out, and stored at room temperature before it is cut. When you do cut the carpet, leave at least three inches of extra carpet in all directions. If you are working with a cut-pile carpet, cut it from the back. Take measurements and use a utility knife, along with a chalk line or straightedge, to make an even cut along the carpet backing. Loop-pile carpet should be cut from the finished side, rather than on the backing.

Carpet installation requires the use of a tool that stretches the carpet. There are knee-kickers and power stretchers available for this part of the job, and both types of tools can be rented at most tool rental centers. Carpet that must be seamed should be seamed before it is stretched. Two pieces of carpet should overlap each other by about one inch at the seam. Make sure the pile of both pieces of carpet is running in the same direction. Use a row-running knife to cut a straight line along the edge of the overlapped carpet. The knife will cut both pieces of carpet simultaneously.

When the cut is complete, remove the cut strip from beneath the top piece of carpet. Lay both edges back to expose the subfloor. Install a strip of hot-melt seaming tape on the subfloor. The tape should be laid so that the center of the tape is in line with the center point of where the two pieces of carpet will meet. Run a hot iron over the seaming tape to activate it. Heat only small sections at a time, and maintain an iron temperature of about 250 degrees F. When the tape becomes sticky, roll the edges of the carpet into place and butt them together. Continue this process, in small sections, until the complete seam is made.

To stretch carpet, you should have both a knee-kicker and a power stretcher. The stretching process is normally started in a corner. Using

231

the knee-kicker, attach the carpet to the tackless strips on two walls, at corners. After the first corner is secured, use a power stretcher to secure the corner directly opposite of the corner already done. Power stretchers have the ability to telescope out to long lengths and can span an entire room. Knee-kickers are used to secure carpet between previously secured locations. Basically, two walls are done with the knee-kicker and two walls are done with the power stretcher. When the carpet is attached to tackless strips, cut away excess carpet with a utility knife. Then use a flat-bit screwdriver to tuck remaining carpet into the gap between the tack strip and the wall.

Carpet at doorways and openings should be cut to size and a metal strip may be installed on top of the carpet. If you are using a metal strip that must be bent over the carpeting, use a wide block of wood and hammer to drive the metal strip down tight. The wooden block should be placed over the strip and tapped down with the hammer. Do not hit the strip with just a hammer, the strip will be damaged. If you are using a nail-on strip, put the metal in place and tack it down.

Selecting carpet

Selecting carpet can be just as difficult as selecting vinyl flooring. There are so many styles, designs, and colors to choose from that customers are often confused. Either take or send your customers to a quality showroom and let the sales staff educate them on the various features and benefits of different carpets. The pad used with carpet is as important as the carpet itself, so don't overlook the attention that should be paid to the pad.

Ceramic tile

Ceramic tile is often found in bathrooms on both floors and walls (Fig. 12-7). If your customer decides to use tile, they will have plenty of choices to choose from. Tile floors can be made of quarry tile (Fig. 12-8), mosaic tile, and glazed ceramic tile. Quarry tile comes in

Figure 12-7

You can do a lot with tile in a bathroom. Courtesy Style-Mark

large squares, and it is produced in natural clay colors. Mosaic tile is small (Fig. 12-9), and generally comes with numerous tiles connected to a single backing. Glazed ceramic tile may be bought as squares or rectangles (Fig. 12-10). You can also offer your customers tile that has designs made in it. The options for tile seem limitless when shopping for just the right look.

Figure 12-8

Quarry tile is an excellent choice for flooring. Courtesy Ackermann

Underlayment should be installed over a subfloor before tile is installed. The underlayment should be at least three-eighths of an inch thick and should be installed with eighth-inch expansion gaps between the sheets. The tile can be secured with adhesives (Fig. 12-11). The adhesive may be organic or epoxy. Epoxy is the preferred adhesive for floors where dampness is a problem. The choice of which type of adhesive to use is often determined by the manufacturer's recommendations. Check with your tile dealer for specifics.

Grout material is a product that fills the gaps between tiles, preventing water and dirt from collecting in the voids. There are numerous types

Figure 12-9

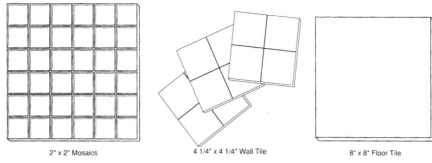

2" x 2" Mosaics   4 1/4" x 4 1/4" Wall Tile   8" x 8" Floor Tile

Notice the types of tiles available to you. Courtesy Dal-Tile

Figure 12-10

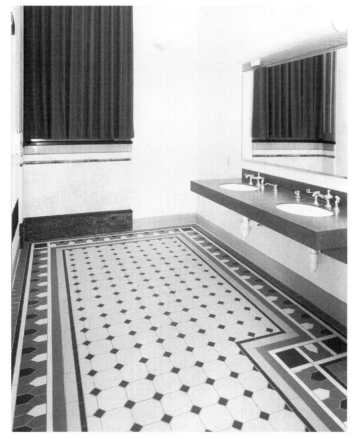

Borders are a popular part of tile installations. Courtesy
Ackermann

235

Figure 12-11

Adhesives for modern tile installations are easy to work with. Courtesy Dal-Tile

of grout material; check with your tile dealer before making a selection. The grout used is important, so don't just buy the first grouting that you come across.

Proper planning is a critical element of good tile installations (Fig 12-12). Deciding how to obtain the proper pattern and spacing will require thought. Special cutters should be used to cut tile (Fig. 12-13). The cutters can be rented at tool rental centers. Installation methods vary, and you should check with your dealer and follow the manufacturer's recommendations for the installation of tile, but let's look at one common way of installing tile.

Before you begin an installation, make sure that you have enough tile to finish the job all at once (Fig. 12-14). Trowel adhesive on the underlayment with a thickness of about one-quarter of an inch. Use plastic spacers (available at tile dealers) to maintain even spacing between your tiles. Lay your first tile in the center of the floor and lay subsequent tiles out from that point. As you set the tiles in the adhesive, press them down firmly. A rubber hammer can be used to tap the tile into place, but it may not be needed. Use a long level to check the consistency of the floor. All tile should be installed level. After the tile is set, you must wait for the adhesive to dry.

Measuring and Setting Perimeter Tile

Measure, cut, and set perimeter tile after all full tiles have been set. Cut and set them one at a time, applying mortar or adhesive as you go.

Figure 12-12

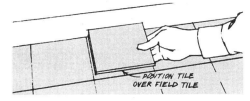

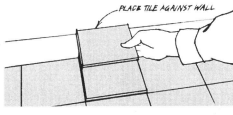

Perimeter tile should be cut after full tiles are in place. Courtesy Armstrong World Industries, Ltd.

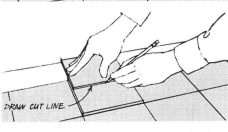

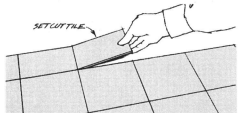

Figure 12-13

IRREGULAR, CURVED, OR CORNER CUTS

Chip away with tile nippers. Take small bites for best results.

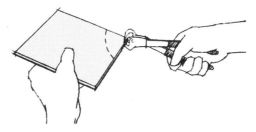

Use a rod saw to cut away marked area.

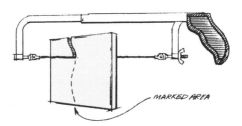

MARKED AREA

Tile can be cut with nippers, rod saws, or electric saws. Courtesy Armstrong World Industries, Ltd.

When the adhesive is dry, you are ready to grout the tile. The grouting should be spread over the floor, filling all gaps between the tiles. This is usually done with a special trowel. Once the grout has filled the cracks, wash the remainder of the grouting off with a wet sponge. To be sure you install the tile properly, follow the tile manufacturer's recommendations.

Tile installation is trickier than it looks. I've done installations myself, but I've really never been quite satisfied with my workmanship. You or your crew might have the skills to do excellent tile work if you don't hire a subcontractor to do the work for you. I said earlier, but I want to emphasize the point, the finished flooring in a bathroom is often a focal point that sets the tone for the entire room. Don't scrimp on getting a professional look on the flooring in bathrooms that you build or remodel. Look over the illustrations here to see what professional workmanship looks like (Figs. 12-15–12-20).

Figure 12-14

SQUARE FEET
OF COVERAGE

NUMBER OF CARTONS	TILE SIZES	
	4¼" x 4¼" 6" x 6"	4" x 4" 8" x 8" 10" x 10" 12" x 12"
	CARTON SIZES	
	12.5 SQ. FT. PER CARTON	11.1 SQ. FT. PER CARTON
1	12	11
2	25	22
3	37	33
4	50	44
5	62	55
6	75	66
7	87	77
8	100	88
9	112	99
10	125	111

A table for tile coverage. Courtesy Armstrong World Industries, Ltd.

Figure 12-15

A fine tile job. Courtesy Ackermann

Figure 12-16

Notice the nice border here. Courtesy Ackermann

Other options

There are other options for bathroom floors, but we have covered the main ones. Some customers will want wood floors in their bathrooms (Fig. 12-21). Personally, I'm against this. Wood floors are not meant to get wet, and bathroom floors often get wet. It is rare to see a bathroom with wood floors, and maybe this is why some customers consider the option. Wood can be an impressive flooring choice (Fig. 12-22). In all my career I have been able to avoid putting wood floors in a bathroom, although I have been asked to on numerous occasions. If you decide to offer wood as an option, you can provide your customer with wood tile in squares, narrow strips of wood flooring, or wide-plank flooring. The selection between these choices would depend on your customers and the theme of the bathroom. Some flooring types are intended for wet areas, and they are the best selection for bathrooms (Fig. 12-23).

There are some other types of flooring that a customer might consider, but none of them are what I would consider appropriate for a modern bathroom. You will have to use your own discretion on how far to go to please a customer. My position on the issue is that my good reputation is riding on the look of my work, and I don't want to tarnish my image by taking on a job that is too far off center from mainstream practices. You make the call and be the judge, but remember that what you do on one job is likely to affect your future jobs, for better or for worse.

Figure 12-17

Large tiles are not common in bathrooms, but large rooms can accommodate them. Courtesy Pella

Figure 12-18

Colored tiles can tie in bathroom themes. Courtesy Ackermann

Editor: Figure caption mentions colored tiles, and the Figure is in b/w.

Figure 12-19

The right flooring and fixtures can make even small bathrooms look luxurious. Courtesy Ackermann

Figure 12-20

Mixing tile colors can provide for interesting flooring options. Courtesy Porcher, Ltd., a Division of American Standard, Inc.

Figure 12-21

Flooring options for bathrooms are limited only by your customer's desires. Courtesy of Andersen Windows, Inc.

▲

Figure 12-22

Wood flooring can be dramatic. Courtesy Andersen Windows, Inc.

Chapter 12

Figure 12-23

Water-resistant wood flooring is best suited for bathrooms. Courtesy Pergo

Walls and ceilings

WALLS and ceilings can account for much of the work required in remodeling a bathroom. If finished walls are not removed, the work may only entail painting, but major remodeling often involves building new walls and recovering old walls. Under these conditions the effort required is considerable, and skills in many trades are needed. You may be faced with framing new interior partitions or the installation of a new door or window. Insulation may need to be added to the exterior walls, and getting a good finish on drywall is something of an art. Even painting is not always as easy as it looks.

Some general contractors and remodelers try to stretch the money that they can make from a job by taking on work that is best left to other trades. Having a rough carpenter attempt to finish drywall may not be a good idea. There are many workers who possess multiple skills. If you have workers who are qualified to do many phases of work, take advantage of your good fortune. However, if your people don't have the skills to finish a job professionally, leave the work to subcontractors. It is the finish work that people see when a job is done, and this can make or break your business.

During my years as a remodeler I've had crews that could do everything from plumbing to tile work. It's not unusual to find workers who can do more than one trade successfully. Ideally, this is the type of employee or piece worker that you should look for. However, getting someone who you think can do the job, but who really can't, can destroy your reputation and your business. Don't turn people loose on multiple tasks until you know that they can do the job right.

I've done some drywall work in the past, but I'm far from an accomplished finisher. Yes, I can do patch work and walls that I've done look okay, but not great. When you are building your business, it's important that all of your work be top-notch. There are plenty of competitors hoping that you will slip up and fall out of competition with them. For some reason, walls and ceilings tend to be targets for wanna-be tradespeople. A lot of people think there is nothing to painting. This simply isn't true. While many people can paint, not a lot of people can produce a paint job that is professional from start to finish. Wallpaper is another phase of work where many people who are not qualified feel that they are. Drywall is a bit more intimidating and fewer people try to fake it in this trade. Ceramic tile looks easy enough to install, but even it has its wrinkles. When you are doing a bathroom remodel, you must be alert to the need for fine finish work.

Framing

Framing work is often called rough carpentry, but that doesn't mean you can use rough estimates or get by with rough skills. If the framing is not done properly, the rest of the job will suffer. For example, if you frame a new wall to hang cabinets on and the wall is out of plumb, you are going to have a tough time hanging the cabinets in a satisfactory manner. Some bathroom remodeling jobs don't require any framing work, but others do. Framing interior partitions is not difficult, but there are a few tricks of the trade that makes the job easier; let's see what they are.

Building a wall

When you are building a wall, you can get the job done in several ways. Most professionals build walls with the framing laying on the subfloor, and then they stand the walls up. This process allows you to frame the entire wall under comfortable circumstances. You may be an accomplished builder or remodeler who is also a carpenter, but many contractors are more business orientated and less familiar with the various trades. For this reason, we will go through some of the basics for those readers who are not well versed in the hands-on part of a job.

Before you start driving nails, lay out the wall locations on the subfloor. Mark the wall location with a chalk line. Once the wall location is known, measure the length of the proposed wall. This will tell you how long your top and bottom (sole) plates should be. Carpenters normally use one 2" × 4" as a bottom plate and two 2" × 4" studs for the top plate of a wall. Begin by cutting the bottom and top plates to the desired length. Next, measure to determine the length needed for vertical studs. Remember to allow for the thickness of your top and bottom plates when measuring between the ceiling joists and subfloor. After you are sure of your measurements, cut the wall studs to the desired length.

Turn the bottom and top plates over on their edges. Place the first wall stud at one end of the plates and nail it into place. Do the same with a second stud at the other end. You have created a rectangle, and all you have to do is install the additional wall studs. Studs are normally installed so that there is sixteen inches from the center of one stud to the center of another. When the wall section is complete, you can stand it up and nail it into place.

If the wall you are building will house a recessed medicine cabinet, a window, or a door, you will have to add rough framing for the device to be added later. Refer to manufacturer specifications for exact rough openings. Don't forget this step, or you will have a harder time later on. Also, if you will be in need of wall blocking, such as would be the case for a wall-hung or pedestal lavatory, put the blocking in while the wall is being built.

Windows and doors

Your framing work may involve framing for windows and doors. It is not uncommon for new windows and doors to be installed during large remodeling jobs. Let's take a look at the types of windows and doors you might want to use and how to frame for them.

Casement windows

Casement windows are well known for their energy-efficient qualities. This style of window offers the advantage of full air flow. When you crank out a casement window, the entire window opens. Casement windows were popular years ago, and now with energy efficiency being more valuable, casement window are once again in the spotlight. While not cheap, these windows perform well in bathrooms and they add a look of distinction to the room (Figs. 13-1 & 13-2).

Figure 13-1

Casement windows are excellent in bathrooms. Courtesy Pella

Figure 13-2

*Casement windows are
available in various sizes
and styles.* Courtesy Andersen Windows

Double-hung windows

Double-hung windows (Fig. 13-3) are the type of windows found in most homes. These windows are generally less expensive than casement windows, and they are well accepted as an industry standard. By far, double-hung windows are used more often than any other type. Bathrooms often require small windows, and double-hung windows are available in small sizes to accommodate all needs in a bathroom remodeling job.

Figure 13-3

A high-quality double-hung window. Courtesy Andersen Windows

Skylights

Skylights and roof windows (Fig. 13-4) can give a bathroom plenty of natural light Rooms filled with sunshine generally appear larger and more inviting. Today's skylights are available with built-in shades and screens. Installing a skylight in a bathroom is a good idea for many reasons. The main reason is additional light without a loss of privacy. If the skylight opens, it can be used to ventilate the bathroom and to reduce the effect of moisture problems. Bubble-type skylights, that don't open, are inexpensive and easy to install, but they don't add to ventilation. When a budget allows for it, a skylight that opens is a good investment.

Figure 13-4

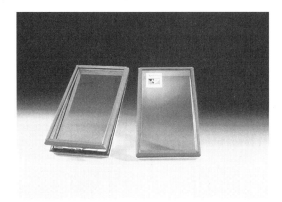

Venting skylights are a good addition to any bathroom. Courtesy Andersen Windows

Figure 13-5

Tilt-out roof windows are ideal for dormer bathrooms. Courtesy Velux-America Inc.

Bathrooms built in dormer areas can benefit from roof windows
(Fig. 13-5). These windows look a lot like skylights, but they are down
lower, where a person can look out of them more like a regular
window. Since roof windows are lower, they can also be opened and
closed, assuming that they are meant to, without the use of a pole.
Skylights that sit high in a ceiling or that are built in light boxes
require the use of an extension pole to operate (Fig. 13-6). Some
poles are turned by hand and others are equipped with small motors
to do the cranking for your customers.

Figure 13-6

Extension poles can be used to open skylights. Courtesy Velux-America, Inc.

Figure 13-7

Fixed-glass panels are an inexpensive way to let a lot of light in. Courtesy Hurd

If roof windows will be used and privacy is an issue, you can offer your customers the option of built-in blinds. The blinds are made inside the glass of the window and work well. They are also attractive. Many of the bathrooms that I've remodeled have had skylights installed over the whirlpool tubs. This is a pleasing way to soak away the aches and pains of a hard day at work. Sitting back in a warm, whirlpool and looking at the stars and moon above is sure to relax most anyone. When you think about bathroom ceilings, consider skylights as an option.

Fixed glass and other options

Fixed glass panels can be used to create walls of glass at an affordable cost (Fig. 13-7). Awning windows (Fig. 13-8) offer some nice design features, and so do circle windows (Fig. 13-9). A number of specialty items are available to give new looks to what would otherwise be plain

Figure 13-8

Consider using awning windows in your next remodeling job. Courtesy Andersen Windows

Figure 13-9

Round and oval windows are good decorating choices. Courtesy Andersen Windows

▲

windows (Figs. 13-10, 13-11, 13-12, & 13-13) A combination of the right windows and accessories can make a new bathroom breathtaking (Fig. 13-14).

Doors

Wood doors cost more than hollow-core doors that are not made of solid wood. They are available in many styles (Fig. 13-15). Some people don't like wood doors because it is possible for them to warp

Figure 13-10

Arches and other designs can complement any job. Courtesy Andersen Windows

Figure 13-11

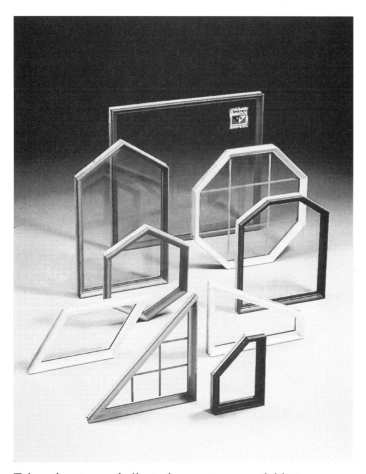

Take advantage of all window options available to you.
Courtesy Andersen Windows

and become hard to operate. This can be especially true in high-moisture areas, such as a bathroom. Hollow-core doors that are not made of wood are less susceptible to warping, but even they can stick. Linen closets in bathrooms can have louvered doors, bi-fold doors, solid wood doors, luan doors, or six-panel hollow-core doors (Fig. 13-16). Choose your doors carefully. Like the walls, ceiling, and windows, a door is a part of the finished product that you produce.

Figure 13-12

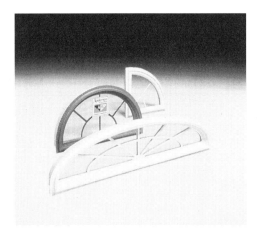

*Adding a new shape to the top of a
window can do a lot for a bathroom.*
Courtesy Andersen Windows

Figure 13-13

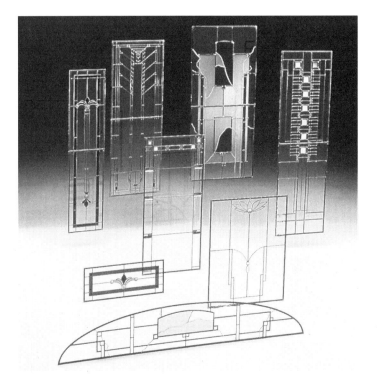

*There are artistic options available for windows in
bathrooms.* Courtesy Andersen Windows

Figure 13-14

Look at the effect windows have on this bathroom.
Courtesy Hurd

Framing window and door openings

Framing window and door openings is simple when building new walls, but it can be complicated if you are cutting new windows or doors into existing walls. Your work will affect the siding on a home and the structural integrity of the exterior wall. The basics of framing window and door openings are about to be explained, but understand

Figure 13-15

Wood doors offer plenty of styles. <small>Courtesy Simpson Door Company</small>

that existing conditions at any home may call for modifications in this phase of the project.

A typical window frame will involve jack studs, cripple studs, and a header. The header will provide strength and support for whatever is sitting on top of the wall. It is usually made with lumber that is nailed together. Jack studs are installed under the header and support it. A horizontal board is installed below the header, at a distance equal to the rough-opening dimension for the window being installed. This board is nailed to the wall studs and supported with short studs, from below. The area above the header is filled with cripple studs. These cripples extend from the header to the top plate, completing the window frame.

Figure 13-16

Six-panel doors are quite popular. Courtesy Simpson Door Company

Framing for a door is similar to framing for windows. Most doors are available as prehung units. These units come to the job ready for installation. The rough door opening should extend all the way to the subfloor. A header, jack studs, and upper cripples will be installed in a manner very similar to window framing. However, the lower framing done with a window is eliminated, and the section of the bottom wall plate that runs through the door opening is cut out.

If you will be cutting into the existing walls of a home for a window or door in the bathroom, look before you cut. Check the other side of your intended opening. Assess what precautions will be needed on the other side of the bathroom wall. Is there some obstruction that will prevent your intended installation? If so, find it before you start cutting. Are electrical wires, heating ducts, or plumbing in the way? You can check this by opening up the interior of the wall, and you should. I've known a number of remodelers who have started cutting

Figure 13-17

Take advantage of window installations anywhere that you can. Courtesy Hurd

before looking ahead and got themselves in trouble. Planning and preparation is a major part of successful remodeling.

Window installation

If you have framed your rough opening properly, window installation is simple (Fig. 13-17). Most windows have nailing flanges for attaching the window unit to the frame walls. Sit the window unit in the rough opening and make sure it is plumb. Nail the unit in place by driving nails through the flange. The flange should be on the exterior side of the house. Always read and abide by manufacturers' recommendations. The instructions are in the box for a reason, and they are not just there for do-it-yourselfers. Far too many skilled tradespeople assume that they don't need to read instructions. After over two decades in the business, I still glance over instructions before starting my work. This may seem silly to experienced workers, but it is just one more way of avoiding problems during the course of a job.

Door installation

When you are working with a prehung door unit (Fig. 13-18), installation is not difficult. Put the unit in place and level it. It may be necessary to install shims around the frame to get it plumb. When the door is plumb, nail the jamb to the framed opening. Regardless of what you are installing, always read and follow the recommendations from the manufacturer of the product. Check to see that the door swings smoothly. If you have to do some minor adjustments, now is the time to do it. Once you know that the door works well, it can be a good idea to remove it from its hinges and set it in a safe spot. All too often doors get damaged during construction and remodeling. If you can protect the finished product by putting it in a safe place, do so.

Insulation

Insulation is not difficult to install, but it can irritate your skin. The easiest type of insulation to install for most remodeling jobs is batt

Figure 13-18

Prehung doors are easy to work with and they look good. Courtesy CraftMaster Marketing

insulation. This type of insulation is available in widths made to fit standard wall and joist cavities. Wall insulation should have a vapor barrier. The barrier should be installed so that it is between the heated room and the insulation. You can buy rolls of batt insulation with a vapor barrier already attached, or you can use unfaced insulation and install sheets of plastic as a vapor barrier. It is important to have the barrier facing the heated room, not the outside of the house. A strong stapler is the only tool needed for installing insulation. Putting the vapor barrier in backwards can result in heavy condensation and moisture damage. There is usually only one wall in a bathroom that requires insulation, and this is a job that most anyone can do. While I recommend leaving certain phases of remodeling and construction to specialized tradespeople, insulation is one aspect of the work that you should be able to rely on your own crews for.

Drywall

Other than for the weight of the material, hanging drywall is not difficult. Finishing it, however, does take some time and practice. When installing new drywall in a bathroom, moisture-resistant drywall should be used. Let's take a look at what is involved with hanging and finishing drywall (Figs. 13-19 & 13-20). Drywall is available in different sizes. Professionals often use 4' × 12' sheets to reduce the number of seams in a job, but 4' × 8' sheets are much easier for the average person to handle. You can choose from different thicknesses to give a finished wall the proper depth. If you will be replacing old, plaster walls with drywall, you will probably have to shim the wall studs to make a consistent surface to hang the drywall on. This can be a real pain in the neck, but sometimes it just has to be done. If the wall studs that you are working with are in good shape, the hanging part of the process is not difficult, and most skilled workers can accomplish the task with a minimum of experience.

Drywall can be hung with the use of nails or screws. Screws are less likely to work loose than nails. If screws are used, an electric screwdriver makes the job go much faster. Screws should be run up tight, to make a depression in the wallboard. When nails are used, they should be driven extra deep, to create a dimple in the drywall. The depressions will be filled with joint compound to hide the nail and screw heads. Don't skimp on nails or screws. Drywall that is not attached firmly can pull loose over time and create flaws in the finished wall.

Drywall can be cut with a drywall saw, jigsaw, or a utility knife. Most pros use utility knives. The procedure requires the drywall to be scored with the utility knife and then broken at the scored seam. You can use a T-square, piece of lumber, or chalk line to make straight cuts.

Hanging drywall on a ceiling is the most difficult part (physically) of any drywall job. Ceilings should be hung before the walls are covered with drywall. When you drywall a ceiling, you will have to make

Figure 13-19

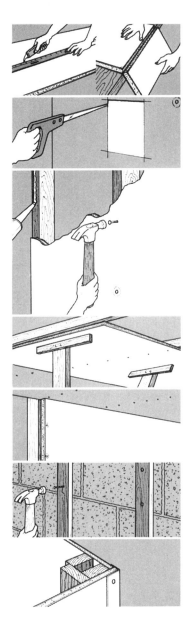

Basic steps to installing drywall. Courtesy Georgia Pacific

Figure 13-20

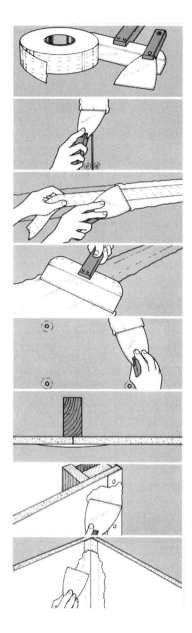

Basic steps to finishing drywall. Courtesy Georgia Pacific

cut-outs for ceiling-mounted electrical boxes, ventilation fans, and so forth. Due to its weight, drywall is not easy to install above your head. However, there is a way to reduce this burden.

A T-brace will be of much assistance when hanging drywall on a ceiling, especially if you or your workers will be working alone. You can make a T-brace from scrap studs. To make the brace, nail a 2" × 4" (about three feet long) onto the end of a 2" × 4" that is long enough to reach the ceiling, with a little left over.

The brace can be wedged under drywall to hold it to the ceiling. The T-arm will rest under the drywall and the long section of the brace will be wedged between the subfloor and the ceiling. It normally takes two people to raise drywall to the ceiling joists, but one person can do it with the aid of ladders and a T-brace. Once a T-brace is wedged into place, it frees one of the people up to attach the drywall to the joists. Two braces can be used to free all hands for other work.

It's possible to hang a ceiling by yourself. Sit the drywall on the tops of two ladders. Leave a couple of feet of the wallboard hanging over each end of the ladders. Put one T-brace under one end of the drywall and raise it with the brace. Wedge the brace against the floor and raise the other end of the drywall with another brace. This will take some time and practice, but once you get the hang of it, you can install a ceiling without help.

Hanging drywall on walls is much easier than hanging it on ceilings. The drywall can be hung vertically or horizontally. If you hang walls vertically you shouldn't need any help. Hanging the drywall horizontally generally results in fewer seams, but it is more difficult to do without a helper. There are, however, some tricks that make horizontal hanging easier for the sole remodeler.

Nail large nails to the studs to provide temporary support for the drywall panel. You can then rest the sheet of drywall on the nails while you attach it to the studs. The large nails can be removed once the drywall is secured. A ledger can be used in place of nails for a more uniform support of the drywall. Nail a 2" × 4" horizontally across the wall studs. Rest the drywall on the ledger while you attach it to the studs. When hanging drywall on the wall studs, don't forget to leave cut-outs for electrical boxes, water supplies, drain arms, and other items that should not be covered up.

Outside corners of walls covered with drywall should be fitted with a metal corner bead. The metal strips protect the exposed corners and edges. These strips are perforated and can be nailed or screwed to wall studs. The corner bead is designed to retain joint compound for a smooth finish. Inside corners do not require a metal corner bead. Drywall tape should be creased and installed to cover the seams of inside corners.

Taping the seams of new drywall is not difficult, but it may take a while to develop a feel for what you are doing. Buy joint compound that is premixed. The tape you will use to cover the seams does not have an adhesive backing; it is held in place by the joint compound. A wide putty knife (about 4" wide) should be used to spread the joint compound over the tape, seams, and dimples.

The first coat of joint compound should be spread over seams, corner bead, and dimples. It is best to work one seam at a time. The first layer of compound should be about three inches wide, and it should be applied generously. Once the compound covers a seam, lay a strip of tape on the compound, and use a putty knife to work the tape down into the joint compound. The tape should sit deeply into the compound. Smooth out the compound and feather it away at the edges of the tape. Continue this process on all seams. Tape is not necessary when filling nail dimples or covering a corner bead. Simply apply joint compound in the depressions until it is flush with the drywall. Smooth the compound out with your putty knife and let it dry.

The first layer of joint compound should dry within 24 hours. A second layer will be applied on top of the first layer, but it should be about twice as wide as the first layer. The second layer must be left to dry, and it should dry within 24 hours. A third layer of compound is usually the final finish on drywall. Before applying the last coat of compound, you must sand the second layer you installed.

Sand the compound first with a medium-grit sandpaper and then with a fine-grit sandpaper. A good dust mask is very helpful during this job.

A sanding block will make the job go faster and will be easier on your hands than using only sheets of sandpaper. Sand the compound with soft strokes to avoid scarring the walls. When the second layer has been sanded properly, you may apply the third layer of compound. This last layer should be about ten inches wide, and the edges should be feathered out. This final layer should be applied in a thin coat.

After the final layer has dried, it must be sanded. Use a fine-grit sandpaper for the finish sanding. When this step is complete, you are ready to clean up and prepare to prime and paint the walls. Some people install baseboards and other trim at this time, and others will wait until cabinets and fixtures are installed.

The installation of interior trim is not difficult, but it does require precise measurements and patience. A miter box and back saw (Fig. 13-21) will be needed for cutting the angles required for interior trim. Once you get the hang of cutting angles, installing trim won't be much of a chore. Baseboard trim should be nailed to wall studs with small finish nails. When baseboard trim meets door casing or a cabinet, it simply butts against the casing or cabinet. Shoe molding is generally installed with baseboard trim when vinyl flooring is used. Shoe molding is small trim that installs in front of baseboard trim. It is often used to cover the joints where vinyl flooring meets a baseboard.

Figure 13-21

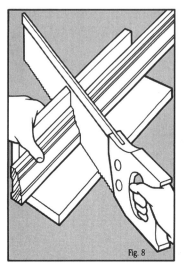

Fig. 8

Miter box and back saw.
Courtesy Georgia Pacific

If the flooring was installed before the baseboard trim, shoe molding is not necessary.

Windows, doors, and open entryways are generally trimmed with casing. The only trick to installing this trim is in cutting the proper angles, and a miter box will make that part of the job nearly foolproof. The nails installed in trim should be countersunk. A nail punch can be used to drive the nail heads deep into the trim. Putty should then be placed in the nail holes before the final paint or stain work is done. If the trim is to be stained, make sure it is made of clear wood and that the putty will not show through the stain. Never use fingerjoint trim when you plan to stain to wood. All of the fingerjoints will show through the stain and look awful.

Paint

Before you begin to paint, vacuum the room to remove all dust. If you don't, your paint will catch the dust, and the job will not look good. You will be working over a subfloor, so dropcloths are not necessary. You must decide whether to use a latex or oil paint. Latex cleans up better than oil, and it will do a fine job on both walls and ceilings. New walls and ceilings should receive at least one coat of primer and one coat of paint. When buying primer, ask the paint dealer to tint it to match the finish color.

Most painters begin their work on the ceilings of a home. Paint rollers work well for applying paint and primer to ceilings. When you paint a ceiling with a roller, you have to cut in along the joints between the walls and ceiling with a brush. Use a two- or three-inch brush to apply a strip of paint to the edges of the ceiling. As soon as the strip of wet paint is applied, lay down the brush and pick up a roller. Use an extension handle in the roller to avoid numerous trips up and down a ladder.

Roll paint on the ceiling and over the strip of fresh paint. Do the cut-in work a little at a time. Trying to cut in the whole ceiling before rolling on the paint can result in a mismatched finish. The cut-in strips will dry before the rest of the paint does. This results in two different

Figure 13-22

An outstanding bathroom ceiling. <small>Courtesy Caradco Wood Windows & Patio Doors</small>

finishes and looks strange. Roll paint on the ceiling in generous amounts; otherwise, it will dry without covering the surface.

With the ceiling finished, you are ready to paint the walls. Apply cut-in strips of fresh paint around the tops of the walls. Follow the same procedures used on the ceiling to avoid mismatched paint. After the first coat of paint or primer you may see imperfections that had been invisible. Take time between the first and second coat to touch up the drywall. Vacuum any dust created from the touch-up work before applying the second coat of paint.

You may wish to texture the ceiling. This is a common option, but some contractors and customers opt for more extensive ceiling work (Figs. 13-22 & 13-23) Assuming that you will texture a ceiling, rather than build a fancy one, there are many options available to you. Joint compound, just like that used to finish drywall, can be used to create a textured ceiling. Some types of paint are sold with a texture made into them. A variety of devices can be used to texture a ceiling. A stiff paint brush can be used to create texture, and a stipple paint roller will also get the job done. Trowels can be used, and even common potato mashers are sometimes used to texture ceilings.

The trim work around windows, doors, and walls will also need to be painted or stained. Paint with a gloss finish is often used on trim when paint with a flat finish is used on walls. Bathroom walls are often painted with a gloss paint; it is easier to clean that flat paint. Before the final paint or stain can be applied to trim work, nail holes

Figure 13-23

A super ceiling with intricate trim. Courtesy Caradco Wood Windows & Patio Doors

must be filled in with putty. Wood putty and a small putty knife is all that is needed for this job. If you will be staining the trim, use a putty that will not show through the stain.

Trim is often stained or painted before it is installed. If you are going to stain the trim, be sure to get clear wood for the trim material. Trim can be stained with either a staining mitt or a brush. After the trim has been stained, you may wish to apply sealer over the stain. This is not a required step, but some people prefer the look and durability offered by sealants.

Other wall options

Other wall options include wallpaper (Fig. 13-24) and tile. There was a time when it was common to find bathrooms with tile floors and tile

Figure 13-24

Wallpaper with borders make attractive bathrooms. _{Courtesy Liz King}

▲

276

Figure 13-25

The use of wallpaper in a bathroom lends a touch of elegance. Courtesy Sanderson

going about half way up the walls. Not many new bathrooms are built this way anymore. The cost of the tile is one reason, but so is cleaning, mildew, and the need for occasional grouting maintenance. Wallpaper (Fig. 13-25) has long been a desirable finish for bathroom walls, and it still is. The use of wallpaper borders (Fig. 13-26) with painted walls is also very popular where walls meet ceilings. Some creative contractors get into using old barn boards or wood planking for interior bathroom walls. Tongue-and-groove planking is also used sometimes. Painted drywall is far and way the most common type of bathroom wall, but there is nothing that says it is the only type of wall covering to use. Creative use of a textured wall can also be very attractive (Fig. 13-27).

Figure 13-26

Stencil-type border is available in a variety of motifs. Courtesy Liz King

There are still customers who will be interested in tile walls. Some of them will want the tile only as a tub or shower surround, and others will want the half-wall look of the past. The customer may want nothing more than a tile border (Fig. 13-28).You or your workers may be able to do the tile work yourself, but it would probably be a good idea to sub the work out to a tile installer if you or your people are not accomplished tile setters. The same goes for wallpaper. Hanging wallpaper looks easy in diagrams, but it is not as simple as it appears. Stenciled borders are easier, but even they can present challenges that are best left to people who work with them all the time.

Using wood as a finished wall in a bathroom can result in problems, due to moisture. If your customer insists on some type of wood walls, make sure the wood is treated to protect it from the heavy moisture it will be subjected to in a bathroom. For most jobs, you should probably stick to drywall and paint or drywall and wallpaper. There is a good chance that you or your workers can perform well with walls and ceilings, but don't attempt to do jobs that you are not confident in your ability to do. Some jobs can be fairly exotic (Figs. 13-29, 13-30, & 13-31).

If you are simply in need of a wall surround for a tub or shower, you can buy factory units that are easy to install (Figs. 13-32 & 13-33). Buy a quality unit. Cheap surrounds are more trouble than they are worth, and you are likely to end up with dissatisfied customers. A good surround will cost a few hundred dollars. Don't be tricked into the inferior models that sell for under $100.

Figure 13-27

Note the textured wall in this bathroom. Courtesy Pella

Figure 13-28

The tile border in this bathroom is a nice touch. <small>Courtesy Pella</small>

Tile

We've talked a little about tile, but let's go into a little more detail on the subject. Bathrooms are natural locations for the use of tile. Be careful not to install too much tile, but be willing to use it for a variety of situations. For example, you can build a shower stall out of tile (Fig. 13-34). A tub surround can be made from tile (Fig. 13-35). Counters to house lavatory bowls can be fashioned out of tile for a distinctive look (Figs. 13-36 & 13-37). You can even use tile to face cabinets or surround for whirlpool tubs (Fig. 13-38).

I believe that tile in most modern bathrooms should be used as an accent. In my opinion, installing too much tile is a mistake. Putting the same type of tile on the floor and walls can be too much for the eyes of most people (Fig. 13-39). In most cases, I would advise against going more than half way up a wall with tile (Fig. 13-40). Of course, my opinions are just that, my opinions. Your customers may love the idea of a fully-tiled bathroom. The use of extensive tile can definitely result in some fantastic bathrooms (Figs. 13-41 & 13-42). Use your imagination and talk with your customers. The right designs and combinations will come to light.

Figure 13-29

Here is a distinctive bathroom. Courtesy Paris Ceramics

Figure 13-30

*Bathrooms with half-wall breaks have
been popular for years.* Courtesy St. Thomas Creations

Figure 13-31

Replica elegance. Courtesy St. Thomas Creations

Figure 13-32

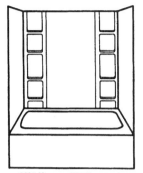

6401 Morocco Wall Surround
- Includes eight shelves.
- High-gloss finish.
- Accommodates window openings.
- Available in U/R colors.
- Comes in six-pack display carton.

6402 Manhattan Wall Surround
- Includes five shelves.
- Color coordinated grab bar.
- High-gloss finish.
- Caulk included.
- Tile-look design.
- Available in U/R colors plus marble tones.
- Can install dome (6692) to complete the bathing area.

6403 Newport

Aqua-Seal°° Wall Surrounds
- Never needs caulking.
- Resists mold and mildew.
- Provides years of watertight beauty and wall protection.
- Handy shelves in reach of tub or shower.
- Can be used with dome top or window trim.
- High-gloss finish.
- Easy to install with step-by-step illustrated instructions.
- Adhesive included.
- Available in U/R colors.
- Can install dome (6692) to complete the bathing area.

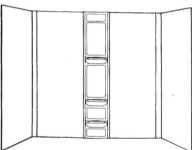

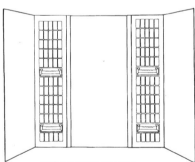

6440 Martinique Wall Surround
- Includes three shelves.
- Sized to fit recessed bathtubs.
- Adjustable sides/backwall.
- Available in white and creme only — matte finish.

6445 Madrid Wall Surround
- Includes four soap and utility shelves.
- High-gloss finish.
- Contemporary tile motif.
- Accommodates window openings.

Basic tub surrounds. Courtesy Universal Rundel

283

Figure 13-33

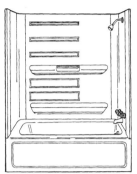

6459 Triumph Wall Surround
- Easy to install three-piece unit.
- Includes three shelves.
- Clear acrylic grab bar.
- High-gloss finish.
- Includes caulk and adhesive.
- Available in U/R colors.
- Can install dome (6692) to complete the bathing area.

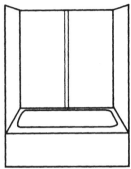

6657 Dover Simplex Wall Surround
- Includes easy-to-maintain texture finish.
- Panel surfaces that resist mold and mildew.
- Complete illustrated do-it-yourself instructions.
- Available in white only.

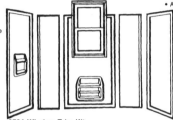

6794 Window Trim Kit
(Wall Surround not included)
- Includes three-piece unit consisting of right and left hand moldings and bottom ledge.

Easy-to-install tub surrounds. Courtesy Universal Rundel

Figure 13-34

A tile shower stall. Courtesy Ackermann

Figure 13-35

A tile tub surround. Courtesy Ackermann

Figure 13-36

Use tile to create a counter surface. Courtesy Ackermann

Figure 13-37

Design distinctive counters with tile.
Courtesy Ackermann

287

Figure 13-38

Make a false front for a whirlpool with tile. Courtesy Ackermann

Figure 13-39

An example of perhaps too much tile. Courtesy Ackermann

Figure 13-40

Running tile halfway up a wall is still acceptable. Courtesy Paris Ceramics

Figure 13-41

Tile is a major part of this bathroom. Courtesy Vista Window Film

Figure 13-42

A bathroom to be proud of. Courtesy Ackermann

Cabinets and counters

CABINETS and counters for bathrooms don't receive nearly as much attention as they do for kitchens. There is no doubt that a kitchen typically contains a lot more cabinets and counter space than a bathroom. However, this does not mean that cabinets and counters should be taken for granted in a bathroom. How many bathrooms have you seen that were equipped with wall cabinets? My guess would be that you have not run across many bathrooms with wall cabinets. Why is this? Is there some law or rule that says a bathroom should not have wall cabinets in it? None that I know of. In fact, wall cabinets are very practical for bathroom use, and they can be quite decorative. A nice, double-door wall cabinet with stained and leaded glass could be a glorious addition to a bathroom. Not only would the homeowner enjoy more storage space, but the beauty of the cabinet would add much to the appearance of the bathroom. So why don't more bathrooms have wall cabinets in them? Probably because they are not a traditional feature in bathrooms. But, maybe it's time to break tradition and start a new way of building better bathrooms.

Base cabinets (Fig. 14-1) have been used in bathrooms for years. The cabinets are called vanities, and they are well accepted fixtures in a bathroom. But again, there are some old-fashioned rules that seem to apply to vanity cabinets. Although, recent years have shown improvements in this aspect of bathroom cabinetry. A simple vanity might consist of little more than a few pieces of particle board put together with a door on the front of the rectangular box. Is this the type of vanity you would want in your home? Probably not, so don't be closed-minded as we explore existing and alternative options for all sorts of cabinets in bathrooms.

If you have as much gray hair as I do, you probably remember well the old, metal medicine cabinets that were considered standard bathroom equipment for so many years. Many of them had a fluorescent light tube either over the top of the mirror or along both sides of it. And of course, there was the old, black electrical outlet made right into the side of the medicine cabinet that was sometimes the only outlet in the bathroom. If you are too young to remember these icons of the past, take my word for it, they existed. The cabinets would rust over time, and the chrome around the mirror would discolor. The slot in the back of the cabinet for spent razor blades always discolored, and I always wondered where all those old,

Figure 14-1

VANITY TOE KICK IS 5½"
VANITY BASE CABINETS HAVE NO
SHELVES

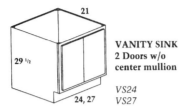

VANITY SINK
2 Doors w/o
center mullion

VS24
VS27

Basic vanity base cabinets.
Courtesy Wellborn Cabinets

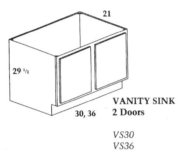

VANITY SINK
2 Doors

VS30
VS36

double-edged razor blades went once they were deposited in the slot. Amazingly enough, some of these antique-like cabinets are still being installed in new bathrooms. When you consider the modern options to the medicine cabinets that time might have forgotten, it's hard to imagine why any contractor would still be installing the old-fashioned metal cabinets.

What other types of cabinets can you remember from old bathrooms? Do you remember the wall-hung lavatories that had legs and a curtain to hides items under the lavatory? I certainly do. How about the old metal cabinets that served as storage for towels and linens? Yeah, there were around for a long time, but their time has passed. We live in a new age, a time when better cabinets are available and should be used.

Standard bathrooms are typically small and starved for storage space. When you can't enlarge the room, you can compensate for storage problems with cabinets. But, you have to get out of the habit of thinking only of a single vanity cabinet. Install a utility cabinet (Fig. 14-2) to increase storage space. Certainly, vanity cabinets are a major source of cabinet and counter space, but there is no reason to stop there. If you look through cabinet brochures or visit a few cabinet showrooms, you can find a wide variety of cabinets that are suitable for use in modern bathrooms. The problem is, too few remodelers and contractors take the time to think about alternative cabinetry. Either this or they don't educate their customers in the options available to them.

As a contractor, you can be limited in what you can do with an existing bathroom. Customers may not be able to afford expansion or an addition beyond the foundation of their homes for bigger bathrooms. One definite way that you can make a fashion statement is with the right cabinets (Fig. 14-3). Now, it is true that good cabinets are not cheap, and this may pose a problem with some budgets. But adding the right cabinets can make a small bathroom much more user-friendly and for a lot less money than an expansion or addition.

We are going to talk about a lot of cabinets in this chapter. I'm also going to provide you with a host of illustrations, so that you can gain a visual picture of what we are discussing. This one chapter alone may change the way you think about bathroom design and remodeling. So, if you are ready, let's get on with the program.

Figure 14-2

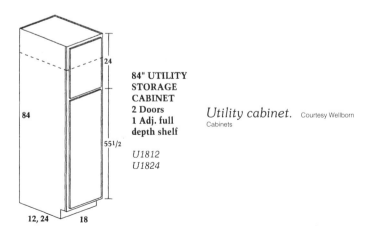

84" UTILITY
STORAGE
CABINET
2 Doors
1 Adj. full
depth shelf

U1812
U1824

Utility cabinet. Courtesy Wellborn Cabinets

Figure 14-3

A picture-perfect bathroom, due largely to the cabinets. <small>Courtesy Merillat Industries</small>

Bathroom size

Bathroom size affects the cabinet choices you and your customers can make. Obviously, a large bathroom can accommodate more cabinets (Fig. 14-4) than a small one can. Ironically, small bathrooms may need more cabinet space than a large bathroom would. One of the first considerations is the amount of closet space, if any, a bathroom will have. Some bathrooms don't have any closet space at all. People get by with such bathrooms, but cabinets offer them an option to a lack of storage, especially if you install wall cabinets (Fig. 14-5). I'm about to take you on a tour of various cabinets. Not all of the cabinets will work in all bathrooms, but you will at least have information here to work with when designing an optimum bathroom space.

Figure 14-4

The combination of light-wood cabinets and a large room makes for a spacious appearance.   Courtesy Merillat Industries

Figure 14-5

- Solid oak fronts.
- European-style self-closing hinges.
- Polished brass finish cabinet hardware.
- High-sheen honey oak finish.
- All exposed surfaces have woodgrain finish.
- Matching vanity, medicine cabinet, and light fixtures available.
- Fully assembled.
- Triple option hardware included to allow choice of brass, matching wood, or china.

Square Raised Panel – 7615183
18″ × 9¾″ × 29¾″ high
Triple option hardware included.

Square Flat Panel – 7616183
18″ × 9¾″ × 29¾″ high
Triple option hardware included.

Wall cabinets for a bathroom.   Courtesy Universal Rundel

297

Base cabinets

Base cabinets are what most people think of when they think of a bathroom. This is because base cabinets have been used as vanities for a long time. There are, of course, a wide range of cabinet styles and qualities to choose from. We will take an item-by-item look at many of your options, but before we do this, I'd like to discuss different aspects of quality that pertain to basically all base cabinets.

What should you look for in a base cabinet? Before you can define your search parameters, you must have a budget in mind. Better cabinets cost more than low-end cabinets. Some inexpensive cabinets look fine, but they may not hold up as well or operate as well as their more expensive cousins. Once you have a price range within which to work, a little time spent with cabinet catalogs or in a cabinet showroom will give you a good idea of what you and your customers are going to get for your money.

I'm looking at a features diagram for a quality base cabinet. The unit I'm looking at has a $\frac{1}{2}$", white, laminated particle board end panel. The depth of the toe kick is $3\frac{1}{8}$", and the particle board toe board is $4\frac{1}{2}$" x $\frac{5}{8}$". This cabinet has a back in it, not all do, and it is a $\frac{1}{8}$" laminated hardboard back. The corner braces for this cabinet are made of plastic and are stapled into the sides and frame of the cabinet. This is a double-door cabinet that has a $3\frac{1}{2}$" x $\frac{3}{4}$" solid hardwood center mullion. The bottom of the cabinet is made of $\frac{1}{2}$" white laminated particle board. One of the nicest features of this cabinet is that the interior shelf height is adjustable, which is somewhat unusual. The shelves are made with edgebanding and dual locking shelf clips. If a drawer is ordered for this style of cabinet, the drawer is rated for a 100-pound holding capacity (Fig. 14-6). This is really good, since so many people stuff drawers full. The drawer slides are epoxy coated, captive, and self-closing. Now let me move to a different grade from the same manufacturer and point out some differences.

This next cabinet is available with end panels of $\frac{1}{2}$" cherry, oak, maple, or hickory wood finished veneer. Both this cabinet and the previous one feature a hanging rail dadoed to receive the cabinet

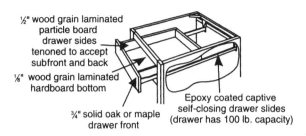

Figure 14-6

½" wood grain laminated
particle board
drawer sides
tenoned to accept
subfront and back

⅛" wood grain laminated
hardboard bottom

¾" solid oak or maple
drawer front

Epoxy coated captive
self-closing drawer slides
(drawer has 100 lb. capacity)

Vanity drawer specifications. Courtesy Wellborn Cabinets

floor. The back on this cabinet is ½" wood grain laminated
hardboard. Other features are about the same. A third model offers
similar features, but with a ⅛" wood grain laminated hardboard back.
The fourth model does not offer adjustable shelves. It too has a ⅛"
wood grain laminated hardboard back.

To look at any of the cabinets I've just discussed you might not notice
much difference. In reality, there isn't a lot of difference. Study
specifications closely (Figs. 14-7, 14-8, & 14-9) to see small
differences. The manufacturer of these cabinets produces good
products, and any of the cabinets would be fine choices. Closer
inspection will turn up certain advantages, such as the adjustable
shelves and some thicker wood or better wood qualities. The types of

Figure 14-7

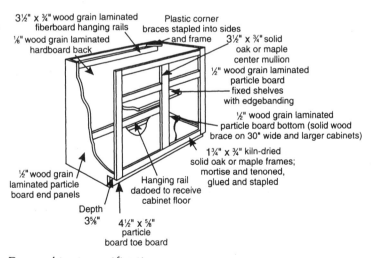

3½" x ¾" wood grain laminated
fiberboard hanging rails

⅛" wood grain laminated
hardboard back

Plastic corner
braces stapled into sides
and frame

3½" x ¾" solid
oak or maple
center mullion

½" wood grain laminated
particle board
fixed shelves
with edgebanding

½" wood grain laminated
particle board bottom (solid wood
brace on 30" wide and larger cabinets)

1¾" x ¾" kiln-dried
solid oak or maple frames;
mortise and tenoned,
glued and stapled

½" wood grain
laminated particle
board end panels

Hanging rail
dadoed to receive
cabinet floor

Depth
3⅝"

4½" x ⅝"
particle
board toe board

Base cabinet specifications. Courtesy Wellborn Cabinets

299

Figure 14-8

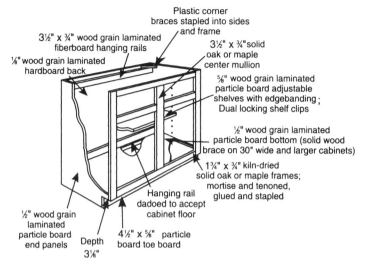

3½" x ¾" wood grain laminated
fiberboard hanging rails

⅛" wood grain laminated
hardboard back

Plastic corner
braces stapled into sides
and frame

3½" x ¾" solid
oak or maple
center mullion

⅝" wood grain laminated
particle board adjustable
shelves with edgebanding ;
Dual locking shelf clips

½" wood grain laminated
particle board bottom (solid wood
brace on 30" wide and larger cabinets)

1¾" x ¾" kiln-dried
solid oak or maple frames;
mortise and tenoned,
glued and stapled

Hanging rail
dadoed to accept
cabinet floor

½" wood grain
laminated
particle board
end panels

Depth
3⅛"

4½" x ⅝" particle
board toe board

Base cabinet specifications.   Courtesy Wellborn Cabinets

Figure 14-9

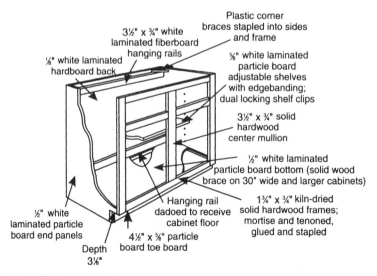

3½" x ¾" white
laminated fiberboard
hanging rails

⅛" white laminated
hardboard back

Plastic corner
braces stapled into sides
and frame

⅝" white laminated
particle board
adjustable shelves
with edgebanding;
dual locking shelf clips

3½" x ¾" solid
hardwood
center mullion

½" white laminated
particle board bottom (solid wood
brace on 30" wide and larger cabinets)

1¾" x ¾" kiln-dried
solid hardwood frames;
mortise and tenoned,
glued and stapled

Hanging rail
dadoed to receive
cabinet floor

½" white
laminated particle
board end panels

Depth
3⅛"

4½" x ⅝" particle
board toe board

Base cabinet specifications.   Courtesy Wellborn Cabinets

doors (Figs. 14-10 & 14-11) could make a big difference to your customer. The point is, you almost have to read cabinet specifications to pick up on the subtle differences. If you were to compare any of these cabinets with the Saturday–Homeowner Specials often seen in newspaper fliers, you would find some major differences. Now that you know that quality can be hidden, let's move onto some specific types of cabinets and their uses in a bathroom. Keep in mind that the concept here is to use a variety of cabinets in conjunction with each other.

Basic vanity base cabinets

Basic vanity base cabinets come in a wide range of sizes and styles (Fig. 14-12). A super-small vanity cabinet may be only 18" x 16" in dimensions, with a height of 19½". This type of cabinet will have only one door and no drawers. It is, to say the least, a minimum vanity size. The next step up might find you using a two-door vanity with dimensions of 24" x 18". Models with widths of 30" or 36" are also available. This type of vanity is the type that is probably used most often. Builders and remodelers often tend to keep vanities simple. Size constraints can also be a factor. To get a small vanity with both drawers and a door requires having only one door. Larger versions of simple vanities might be 24" x 21" or 27" x 21". Sizes continue up to a width of 36". This type of selection is what your basic vanity base cabinet is likely to be like.

Vanities with drawers

Vanities with drawers are very nice (Fig. 14-13). To get this type of arrangement, you have to either have space for a larger vanity or you will have to sacrifice one door to have the drawers. Many people are willing to live with a single door if they get drawers as a replacement for the second door. A typical vanity that has one door and three drawers could have dimensions of 24" x 21", 30" x 21", or 36" x 21". If you want a vanity base that has two doors, two drawers, and a drawer blank (Fig. 14-14) (dummy drawer), the size will start around 36" x 21". Other sizes could be 42" x 21" or 48" x 21".

Figure 14-10

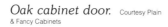

Beaded inset cabinet door.
Courtesy Plain & Fancy Cabinets

Figure 14-11

Oak cabinet door. Courtesy Plain
& Fancy Cabinets

Figure 14-12

2 doors, 4 drawers – 7616483 – 48″ × 18″ × 29¾″ high
2 doors, 4 drawers – 7816483 – 48″ × 21″ × 29¾″ high
Triple option hardware included.

2 doors, 2 drawers – 7616303 – 30″ × 18″ × 29¾″ high
2 doors, 2 drawers – 7616363 – 36″ × 18″ × 29¾″ high
2 doors, 2 drawers – 7816303 – 30″ × 21″ × 29¾″ high
2 doors, 2 drawers – 7816363 – 36″ × 21″ × 29¾″ high
Triple option hardware included.

1 door, 2 drawers – 7616243
24″ × 18″ × 29¾″ high
1 door, 2 drawers – 7816243
24″ × 21″ × 29¾″ high
Triple option hardware included.

2 doors – 7613243 – 24″ × 18″ × 29¾″ high
Triple option hardware included.

A selection of vanity base cabinets.   Courtesy Universal Rundel

Figure 14-13

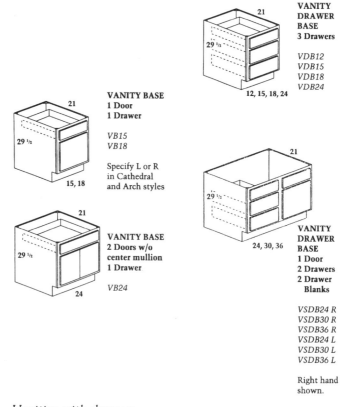

VANITY
DRAWER
BASE
3 Drawers

VDB12
VDB15
VDB18
VDB24

12, 15, 18, 24

VANITY BASE
1 Door
1 Drawer

VB15
VB18

Specify L or R
in Cathedral
and Arch styles

VANITY BASE
2 Doors w/o
center mullion
1 Drawer

VB24

VANITY
DRAWER
BASE
1 Door
2 Drawers
2 Drawer
  Blanks

VSDB24 R
VSDB30 R
VSDB36 R
VSDB24 L
VSDB30 L
VSDB36 L

Right hand
shown.

Vanities with drawers. Courtesy Wellborn Cabinets

More elaborate vanities with drawers could have one door, six drawers, and one drawer blank in a size of 42" x 21". This is a lot of vanity for the space taken up by it. To get six drawers, one drawer blank, and two doors, you would step up to dimensions of 48" x 21". Now you are getting into a really spacious vanity.

Side cabinets

Once you have your basic vanity base cabinet, you can add side cabinets, if space allows. For example, you could add a one-door, one-drawer side cabinet if you have an extra 15 inches of width. A double-door side cabinet with a full-width drawer is available in a 24" package. A three-drawer base cabinet will add only 12 inches

Figure 14-14

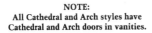

NOTE:
All Cathedral and Arch styles have
Cathedral and Arch doors in vanities.

VANITY TOE KICK HEIGHT IS 5½"

VANITY BASE CABINETS HAVE NO
SHELVES

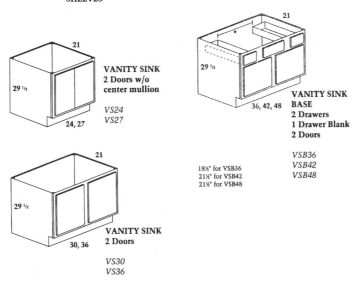

21

29 ½

VANITY SINK
2 Doors w/o
center mullion

VS24
24, 27   VS27

21

29 ½

36, 42, 48

VANITY SINK
BASE
2 Drawers
1 Drawer Blank
2 Doors

VSB36
VSB42
VSB48

18½" for VSB36
21¼" for VSB42
21½" for VSB48

21

29 ½

VANITY SINK
2 Doors
30, 36

VS30
VS36

Vanities with double doors and drawer blanks. Courtesy Wellborn Cabinets

to your overall width (Fig. 14-15). Wider models are available.
There is even a vanity hamper cabinet that offers a tilt-down door
that will install in a cabinet. The width of the hamper is 18 inches.
If you are thinking about using kitchen base cabinets in a
bathroom, you should know that the height for kitchen cabinets
is more than it is for bath cabinets. A standard height for a bath
cabinet is 29½", while a similar kitchen base cabinet will stand
34½" tall.

Vanity linen cabinets

Vanity linen cabinets can take on any size (Fig 14-16). You can even
use a kitchen pantry cabinet as a linen cabinet if you want to. A
narrow one, say about 18 inches wide and about 21 inches deep will

Figure 14-15

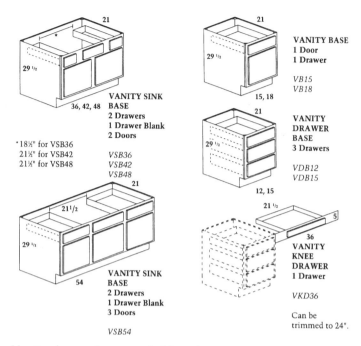

Vanity drawer bases and side cabinets. Courtesy Wellborn Cabinets

provide your customer with three drawers and a door that is 55 inches high. If you have room for one that is 24 inches wide, you will get double doors. When space allows for such a cabinet, the increased storage area is substantial. Cabinet finish on the linen cabinets can be ordered to match the other cabinets that you install in the bathroom. If you have cabinet-grade carpenters working for you or with you, it's always possible to custom build a similar type of cabinet to meet nearly any space limitations. I've had my carpenters build a number of such cabinets in past years.

Toilet and vanity wall cabinets

Toilet cabinets are made to hang over a toilet (Fig. 14-17). One particular brand that I am familiar with has two doors with two adjustable shelves. The unit measures 24" x 36" and is an attractive addition to a bathroom. With a depth of eight inches, this cabinet is suitable for many uses. The same company makes what they call a

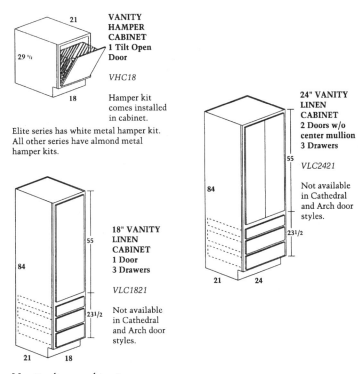

Figure 14-16

VANITY
HAMPER
CABINET
1 Tilt Open
Door

VHC18

Hamper kit
comes installed
in cabinet.

Elite series has white metal hamper kit.
All other series have almond metal
hamper kits.

24" VANITY
LINEN
CABINET
2 Doors w/o
center mullion
3 Drawers

VLC2421

Not available
in Cathedral
and Arch door
styles.

18" VANITY
LINEN
CABINET
1 Door
3 Drawers

VLC1821

Not available
in Cathedral
and Arch door
styles.

Vanity linen closets.   Courtesy Wellborn Cabinets

vanity wall cabinet. This cabinet is twelve inches wide, thirty inches
tall, and four inches in depth. It has one door and three adjustable
shelves. One somewhat unique aspect of this cabinet is that it can be
mounted in a recessed position or used as a flush mount unit. The
two cabinets that I've just described are only a fraction of what is
available to you and your customers.

If you have some wall space to spare in a bathroom but can't afford
a deep linen closet, you could consider a semi-custom utility
cabinet. One manufacturer that I'm familiar with offers just such a
creature. The cabinet stands 84 inches tall and has a depth of eight
inches. This is plenty of depth for many bathroom storage items,
and the large size of the cabinet can accommodate plenty of stuff.
This type of cabinet costs a good deal more than its full-size
counterpart, but it's hard to beat for extensive storage in a package
with a low profile.

Figure 14-17

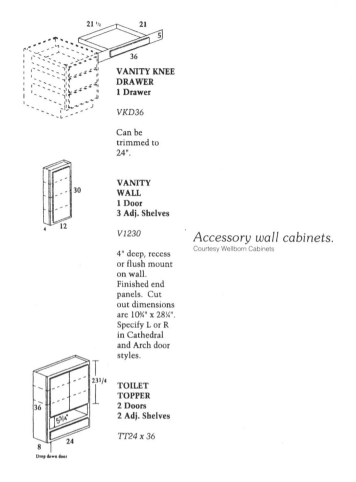

VANITY KNEE
DRAWER
1 Drawer

VKD36

Can be
trimmed to
24".

VANITY
WALL
1 Door
3 Adj. Shelves

V1230

4" deep, recess
or flush mount
on wall.
Finished end
panels. Cut
out dimensions
are 10¾" x 28¼".
Specify L or R
in Cathedral
and Arch door
styles.

TOILET
TOPPER
2 Doors
2 Adj. Shelves

TT24 x 36

Accessory wall cabinets.
Courtesy Wellborn Cabinets

Regular wall cabinets

Regular wall cabinets, like those used in kitchens can be used in
bathrooms. One problem that you might run into is the standard
depth of a kitchen cabinet. The normal depth for these cabinets is
often 24 inches. However, models are available with depths of 12 or
18 inches, which gives you a lot of versatility in a bathroom setting
(Fig. 14-18). For example, one manufacturer offers a three-door wall
cabinet that is 12 inches high, 12 inches deep, and 48 inches long. It
is just the right size for fitting over a nice vanity, and it adds a lot of
storage space. Under-cabinet lighting can be used to illuminate the
vanity top, and track lighting could be mounted on top of the cabinet

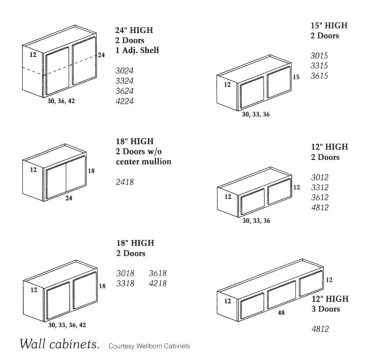

Figure 14-18

24" HIGH
2 Doors
1 Adj. Shelf

3024
3324
3624
4224

30, 36, 42

15" HIGH
2 Doors

3015
3315
3615

30, 33, 36

18" HIGH
2 Doors w/o
center mullion

2418

12" HIGH
2 Doors

3012
3312
3612
4812

30, 33, 36

18" HIGH
2 Doors

3018   3618
3318   4218

30, 33, 36, 42

12" HIGH
3 Doors

4812

Wall cabinets. Courtesy Wellborn Cabinets

to provide full-room illumination. There is also a two-door model that comes in widths of 30, 33, and 36 inches.

Are you having trouble thinking of a bathroom with overhead cabinets in it? That's okay, it's not a common concept, but there is no reason why you and your customers should not take advantage of available space. A wall cabinet is an ideal solution to storage when space is very limited. Fitted with the right door fronts, the cabinet can be a major enhancement to the bathroom.

Medicine cabinets

Medicine cabinets have long been an expected amenity in a bathroom. Many modern bathrooms don't contain medicine cabinets. Mirrors have replaced some medicine cabinets. This is okay, but a medicine cabinet does provide storage and a mirror, so why not take advantage of the storage. A simple medicine (Figs. 14-19 & 14-20) cabinet might measure about 18" x 30", with a depth of about 4 ´

Figure 14-19

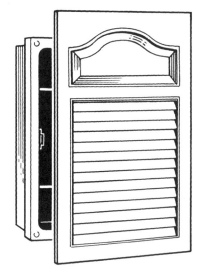

A simple, louvered medicine cabinet. Courtesy Nutone

AH-70N American Heritage

Figure 14-20

Woodgrain medicine cabinet.
Courtesy Nutone

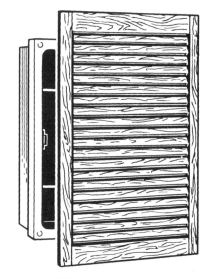

AH-75 American Heritage

inches. In a cabinet this size you would likely get two adjustable shelves. A cabinet like this is standard issue, so to speak. A more elegant unit would have the same height and depth, but may have a width of anywhere from two to four feet. In this case, you would get three doors and six adjustable shelves (Fig. 14-21). The options for

Figure 14-21

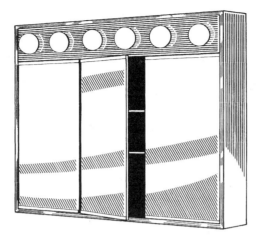

D-4210-G/S/W Europa TriVista

Three-door medicine cabinet. Courtesy Nutone

medicine cabinets are extensive. Rather than go into a lot of talk about the many different types, I will provide you with some illustrations to consider (Figs. 14-22 & 14-23).

The public sometimes has a bad taste in their mouths due to the old-style medicine cabinets that they grew up with. Remember early in the chapter, when I described the type of medicine cabinet I remembered from younger years? Well, probably a lot of people associate all medicine cabinets with those old beasts. New styles are very different and offer a lot of eye appeal, not to mention storage. I think that if you take the time to show your customers some examples of modern medicine cabinets that they may change their outlook on the subject.

Choosing base cabinets

Before you begin choosing base cabinets, you owe it to yourself to shop around. You may be amazed at the number of variations available in base cabinets. You will have decisions to make on sizes, styles, colors, and features. Cabinets are fundamental elements of many bathrooms. The cabinets may attract more attention than any other feature in the room. They will also receive a lot of use. Since

▲

311

Figure 14-22

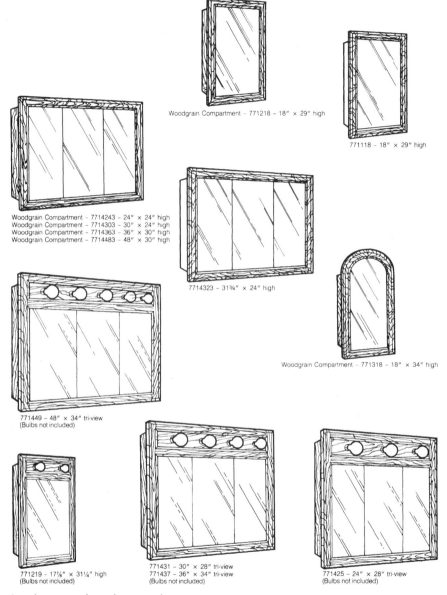

Woodgrain Compartment – 771218 – 18″ × 29″ high

771118 – 18″ × 29″ high

Woodgrain Compartment – 7714243 – 24″ × 24″ high
Woodgrain Compartment – 7714303 – 30″ × 24″ high
Woodgrain Compartment – 7714363 – 36″ × 30″ high
Woodgrain Compartment – 7714483 – 48″ × 30″ high

7714323 – 31¾″ × 24″ high

Woodgrain Compartment – 771318 – 18″ × 34″ high

771449 – 48″ × 34″ tri-view
(Bulbs not included)

771219 – 17⅞″ × 31¼″ high
(Bulbs not included)

771431 – 30″ × 28″ tri-view
771437 – 36″ × 34″ tri-view
(Bulbs not included)

771425 – 24″ × 28″ tri-view
(Bulbs not included)

A selection of medicine cabinets. Courtesy Nutone

Figure 14-23

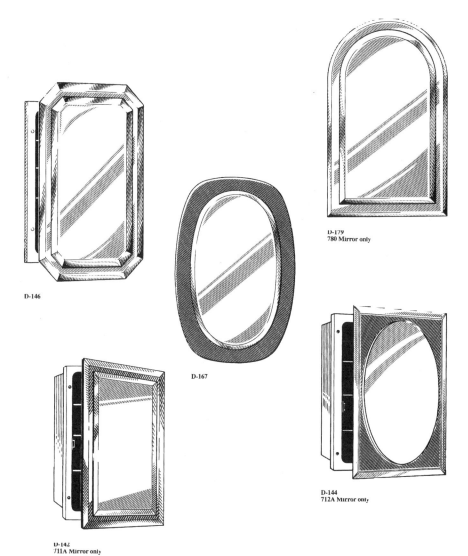

D-146

D-167

D-179
780 Mirror only

D-142
711A Mirror only

D-144
712A Mirror only

Mirror-type medicine cabinets. Courtesy Nutone

you will want your job to be beautiful, functional, and enjoyable to live with, take your time in helping your customers choose their cabinets.

Base cabinets are available in different widths, as you have already seen. The base cabinet that will sit under the lavatory will have an open space for the lavatory bowl to sit in. The lavatory base can range in width from about 18 inches to six feet, or so. Other base cabinets may be as narrow as one foot or as wide as four feet, or more. There are, of course, other sizes available, and custom cabinets can be made to your customer's specifications.

Custom cabinets are generally much more expensive than production cabinets. With the wide selection of production cabinets available, there is rarely a need for custom cabinets. Some people want their cabinets built just for them, but most people will have no trouble finding stock cabinets to suit their needs and desires.

Cabinet materials can consist of solid wood, plywood, and particle board, or pressed board. Many production cabinets use a mixture of these materials. Deciding on what the cabinets are made of is only part of the buying decision for your customers. They will also have to look at the construction features of the cabinets. For example, Dovetail joints should last longer than butt joints.

Other considerations for choosing base cabinets include whether the cabinet will have doors, drawers, or special accessories (Fig. 14-24). An important consideration in choosing a drawer base is how well the drawers glide. Insist on a cabinet with quality glides and rollers.

Choosing wall cabinets

Choosing wall cabinets will be similar to choosing base cabinets. Your customers will have to consider the sizes and styles that best suit their requirements. Will they want cabinets with glass doors, raised-panel doors, doors with porcelain pulls, or door with finger grooves? There are plenty of choices to contemplate with cabinets. The heights and

Figure 14-24

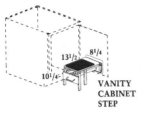

VANITY CABINET STEP

C2713

Folds up to 3⅜" thickness.

Vanity cabinet step is made of ¾" solid wood with ¼" poplar hardwood top. Fits in all vanity cabinets.

Special vanity accessories.
Courtesy Wellborn Cabinets

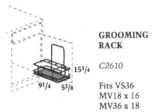

GROOMING RACK

C2610

Fits VS36
MV18 x 16
MV36 x 18

Fits as small as 12" wide openings. Rack mounts to door. Provides storage for one hand-held hair dryer, two cans of hair spray and two curling irons. White metal.

widths of wall cabinets vary, and you can offer your customers sizes to fit their needs.

Your customers should look for quality in the shelves and latches of wall cabinets. The supports for shelves should be adjustable and allow random spacing of the shelves. Magnetic latches are usually favored over plastic latches. Let your customers inspect hinges, structural supports, and all other structural aspects of wall cabinets before they buy them.

Bathroom counters

Bathroom counters can be made of many materials. Some remodelers and builders like to use the same types of counters used in kitchens. Others prefer molded tops with integral lavatory bowls. The molded tops are my preference. Some contractors like to make their own tops, and others install tops that are finished with ceramic tile. The bottom line is what the customer wants. I've installed quite a few tile tops in custom bathrooms, and they have looked very nice. The preformed tops that you see in kitchens are used a lot, but I have seldom used them. Cultured-marble tops have gone into a large number of my bathroom jobs. They are, by far, the predominant top in my bathroom work. High-fashion molded tops have also seen a fair amount of use in my bathroom remodeling and construction.

I like the designer tops with their special features. For example, one manufacturer offers a top that is so tough that burns from cigarettes can literally be sanded out of the finished surface. I don't smoke, but I can see where this could be a nice feature in some homes. For the money, I like cultured-marble tops the best. To me, a kitchen counter with a drop-in lavatory looks cheap. This is just my personal opinion, and I know a lot of people will disagree with my statement, but it is the way I feel.

When your customers shop for a counter, they are likely to be confused by all the options. Be prepared to help them with explanations and advice. Most bathrooms have relatively little counter space in them, so a high price by the linear foot is not as discouraging as it would be in a kitchen job. There are some very interesting styles and patterns available to your customers, and it may take them a while to decide. Be patient with them. Make sure that the top they select is available for delivery within your production schedule. Some tops are special-order only, and they can take quite a long time to arrive. Don't let the job be held up if you can avoid it. Your customers need their new bathroom, and you need your cash flow.

Cabinet installation

Cabinet installation should begin with a design that has been studied and approved. It is much easier to make changes in a cabinet design on a drafting table or computer than it is on the job. You should have a good cabinet layout drawn before you begin the installation process. Any good cabinet supplier will provide recommended designs and drawings. When you are ready to install cabinets, follow the design, but before you jump right into setting and hanging cabinets, double-check your previous work.

Check the floor and walls to make sure they are plumb and level. Cabinets that are not installed level may not operate properly, and there may be visual evidence of the poor installation. Shims can be used to overcome minor problems with walls and floors, but you should know what you are dealing with before you begin installing cabinets.

Install wall cabinets

Install wall cabinets first. There may not be many, if any in a bathroom, but it is a good idea to install them first. By installing the wall cabinets first, you reduce the risk of damaging base cabinets, and you will have more freedom of movement for the job.

There are no rules to indicate how high you must hang wall cabinets. Let your customers tell you what a comfortable height is for them. Once you determine the desired height for the cabinets, mark a level line as a reference point. Before you put the cabinets in place, find the wall studs. If you have gutted the bathroom and installed new drywall, you will be familiar with the stud locations. You may have even thought ahead and marked their location on your remodeling plan for easy reference. If you have trouble locating the studs, don't hesitate to probe the wall where the cabinets will be hung. The back of the cabinet will conceal any holes you make in the wall.

Cabinet installation is much easier when at least two people are working on the job. Many professionals place props under the cabinets they are hanging to help keep the units in place prior to permanent attachment. The props can be made from lumber you have left over from the rough carpentry work. Hanging wall cabinets in a bathroom is a little different than hanging them in a kitchen. There are fewer cabinets, less weight, and the cabinets are generally smaller and easier to work with.

When the first cabinet is in place and level, drill holes through the back of the cabinet and into the wall studs. The holes should be kept near the top of the cabinet. Most cabinets have mounting strips for the screws to penetrate. Install screws to hold the cabinet in place. Check the unit to make sure it is level and plumb. If it isn't, use shims to get the cabinet installed properly.

With the first cabinet installed, you are ready to install adjacent cabinets, if there will be any and there probably will not be, in the same manner. Adjacent cabinets should be attached to each other. Make sure the cabinets are aligned uniformly. Use small screws to attach the two cabinets to each other. The screws should be installed near the top and bottom of the cabinet side walls.

If you have help available, you may want to put the cabinets together on the floor and then raise them to the wall as one unit. Take your time in aligning the cabinets. When the cabinets are all attached, simply raise them to the wall and support them with prop sticks. Position the cabinets so that they are level and plumb, and then screw them to the wall studs. If you have chosen to install a valance, do it before you set the base cabinets.

Installing base cabinets

When installing base cabinets, start with the lavatory base cabinet and build out from it with remaining cabinets. Base cabinets should be attached to each other in the same way as wall cabinets. Check frequently to see that the base cabinets are level and plumb. It may be necessary to shim under the cabinets to keep them level.

Not all base cabinets have sides and backs: some are just fronts. This type of unit requires the installation of cleats. Cleats are just strips of wood that support the countertop. The cleats should be attached to wall studs with the top of the cleat at the same height as the tops of adjacent cabinets.

Install the counter

Once you have your base cabinets set, you will want to install the counter. Some people wait until the base cabinets are in to order their counters. Working in this manner slows down the progression of the job, but it eliminates much of the risk of getting a counter that is not sized properly. This isn't a bad idea for kitchen work, but it is rarely necessary for bathroom tops. The counters in bathrooms are usually straight, without angles and turns, so you can feel fairly safe in ordering your tops in advance. If you are buying your cabinets and counter from a good supplier, your cabinet layout was probably drawn well in advance, and it is likely you already have the counter. Assuming that you have your counter, let's see how it should be installed. For this example, we are talking about the same types of counter that would be used in a kitchen, where a drop-in lavatory bowl will be used. Cultured-marble tops will install differently, and we will talk about them soon.

Look down on your base cabinets; you should see some triangular blocks of wood or plastic in the corners. These triangles provide a place to attach the counter to the cabinet. Before sitting the counter in place, drill holes through these mounting blocks. Keep the holes in a location that will allow you to install screws from inside the cabinet. You may want to drill the holes on an angle, towards the center of the cabinet. This will make the installation of screws easier.

Position the counter on the base cabinets, and check its fit. When you are satisfied with the positioning of the top, install screws from below. The screws used should be long enough to penetrate the triangular blocks and the bottom of the counter, but be certain they are not long enough to come through the surface of the counter and ruin it.

Bathroom counters must have a hole cut in them for the lavatory bowl. The supplier of the top will often cut these holes if they are provided with information on the size and location of the lavatory, but you can cut your own hole on site. If you cut your own sink hole, use the template that came with your new lavatory. If you don't have a template to work with, turn the lavatory upside down and sit it in place on the counter.

Lightly trace around the lavatory rim of a self-rimming lavatory with a pencil. Remove the lavatory and draw a new outline inside the original tracing. The hole must be smaller than the lines you traced around the lavatory. There must be enough counter left after the hole is cut to support the rim of the fixture.

When you are ready to cut out the hole, drill a hole in the counter, within the perimeter of the lavatory outline. Use a jig saw to cut out the hole. Put the blade in the hole you drilled and slowly cut the hole. Remember, the hole you make must be smaller than the outline of the lavatory. After the hole is cut, set the lavatory in it and check the fit; you may have to enlarge the hole a little at a time to get a perfect fit. Once the lavatory fits well, you can proceed with the plumbing connections.

Premolded tops, like cultured-marble tops have the lavatory bowls built into them. These tops are heavy enough to just sit on the base cabinets. There is no need to attach them to the cabinets with screws. In fact, don't attempt to do this, because you will probably damage the top. The weight of the top, along with caulking and plumbing connections will hold it in place. This is the typical installation procedure for any type of lavatory top that comes to you premolded.

Now that we have completed the basics of cabinet selection and installation, we are ready to move onto the last chapter. There we will discuss hardware and accessories that will put the final finishing touches on your work of art in the remodeling field. Let's turn to Chapter 15 now and see what goodies we can find.

15

Accessories to enhance a bathroom

ACCESSORIES to enhance a bathroom are a subject that many builders and remodelers take for granted. A lot of contractors assume that their customers are going to take care of the accessories for a bathroom. Most homeowners think that the cost and installation of accessories will be included in their total job price. Which way will it be? Make sure that you and your customers have a clear understanding of who is doing what for whom and at what price the work is being done for. About the last thing that you want is to have a job that has gone smoothly right to the end wind up in a conflict over bathroom accessories.

What should be considered a bathroom accessory? Ah, now that is a good question. Typical accessories include a holder for toilet tissue, a towel rack, and a toothbrush holder. From this point, anything could happen. For example, who is providing the light bulbs for the light fixtures? When I built my first house, I thought the electrician would be providing the light bulbs. I was wrong. The electrician had bid the job for labor and material, which included light fixtures, but not light bulbs. This may not seem like a big deal, but when you add up all the light bulbs in a large house and multiply it by several houses a year, the cost is worth taking note of. So, who is going to pay for the light bulbs that will go into the bathroom job you are just finishing? I know that light bulbs are a stretch in terms of calling them accessories, but someone has to buy and install them.

I've done a lot of bathroom jobs, and I've bought and installed a lot of accessories. Many of the accessories have been common, stock items that were available from most any store selling bathroom products. The most common selection has probably been an oak set of accessories that includes a soap dish, a toothbrush holder, a bath towel bar, a towel ring, and a toilet-tissue holder. For whatever reason, this type of package deal seems to appeal to people, myself included. Yes, I have used the same set in homes of my own. The type of set that comes in second behind the oak package eliminates the towel ring and is offered in a chrome finish. Both of these packaged deals are affordable and provide the basics for bathroom accessories. But, there are a lot of other options available.

Let's go over a little list of items that might be considered bathroom accessories. If a bathroom is equipped for showers, it should have either a shower rod and curtain or shower doors. These seem like accessories to me. Even bathtubs without showerheads sometimes have sliding doors for privacy. Holders for toilet tissue are considered essential, as are towel bars or rings. Toothbrush holders are common, and soap dishes can be nice. What else is needed or would qualify as an accessory? You may not think of some of the following items as something that a contractor would install, but it depends upon how far you take your jobs. For example, bath mats, toilet rugs, and items of this nature might fall into your job description. These types of items, however, are usually provided by the homeowner. What about the window treatments? Are you responsible for blinds, shades, or curtains? If you are not providing any of these actual items, are you supposed to hang rods or bars for them? Talk to your customers and define where your responsibilities end, so that there will be no confusion or hard feelings as you complete your job.

Some contractors feel that once the main remodeling is done, their job is done. This might be the case, but if your customers don't understand where your work begins and ends, you can find yourself in difficult situations. I would agree that a remodeler would not normally provide and install bath mats and curtains. However, there are remodeling and building firms that take their jobs all the way through the decorating stages, right down to waste baskets. If you run into a customer who has talked to such a company, the customer may assume that your company is planning on the same type of work. I can't stress enough that you must draw a clear line for where your work will end.

I've been remodeling bathrooms and building homes for a long time. It never seems to fail that most problems with customers crop up in the last ten percent of the job. What a waste! All that good effort and relationship that has gotten you 90 percent of the way can be blown out of the water right at the end. Don't let this happen. I don't want to preach to you here, we are supposed to be talking about accessories, but it is important to the health of your business for you to know what you might be getting yourself into as you near the completion of a job.

Shower rods

Shower rods are a bathroom component that most customers expect contractors to install. It is often the plumbing contractor who installs shower rods. A majority of the rods are simple, chrome tubes that span a distance and that are held in place by two receiving brackets. Shower rods are easy to install. But, are you supposed to provide and install the shower curtain, as well? Probably not, but you have to make this arrangement with your customer. If you are going to get involved with shower curtains, there are a lot of styles, colors, and patterns for your customers to choose from. I suggest that you leave the selection and hanging of a shower curtain to your customers.

Shower doors and tub enclosures

Glass shower doors and tub enclosures are a definite upgrade over shower curtains. Some people don't like glass doors, because of the soap scum that will build up on them. A few parents worry that their children will break the glass enclosure and be injured. Most people don't like shower curtains, because they mildew easily. How can you win? The best way, in my opinion, is to lay low on advice and let customers make up their own minds. In terms of job appearance, a glass enclosure is much more elegant. Glass doors are also more effective in retaining water, to avoid problems with leaking floors. Assuming that your customer wants a glass enclosure, should you or your people install it, or should you sub the job out? This depends on you and your crews. I have installed a number of bath and shower enclosures. They are not too bad to install in new work, but they are a serious pain to install in existing bathrooms where walls and floors are not square and plumb. Personally, I sub this phase of my jobs out to glass companies.

If you want to get a bit creative in the tub or shower enclosures, you might consider building part of the bathing surround out of glass blocks. These blocks provide some light, but they break up images well enough to be used for privacy enclosures. The cost of using blocks may be more, but it adds a different look to the bathroom.

Figure 15-1

Beamed ceilings are a nice feature in some bathrooms.
Courtesy of Weathershield

Something as simple as this could slowly become your remodeling signature in bathroom work. Always look for something special that you can do on most jobs that will be similar to your remodeling trademark.

False beams

Adding real, or false, beams to a bathroom ceiling (Fig. 15-1) can produce interesting results if the style of the bathroom will accept the decoration. Beams can be used with flat ceilings, but they look better on vaulted ceilings. There are various sizes and types of false beams available. Obviously, this type of accessory is not for every bathroom, it is not even suitable for most bathrooms. But, there are times when replica fixtures are used that beam ceilings can be a nice touch. Keep an open mind about bath accessories, you don't have to be limited to soap dishes.

Holders for toilet tissue

Holders for toilet tissue (Fig. 15-2) are a must in most bathrooms. There are two basic types. You can buy the type that is flush-mounted on a wall or the type that is recessed into a wall or vanity cabinet. The

Figure 15-2

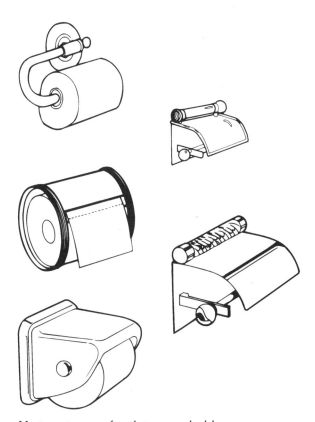

Various types of toilet-paper holders. Courtesy of Eljer

recessed models are nice in that they don't protrude into the usable space of the bathroom. I like the recessed models from this viewpoint, but I must admit, flush-mount units are what I install the most of. Chrome and oak are the two types of finishes that I get the most demand for. There are probably dozens of designer models of tissue holders that your customers may be interested in. I can tell you that some people don't like recessed holders mounted in the sides of vanities. In my earlier years, customers would ask to have the holders mounted in the vanities to save space and then later complained about the space lost in the vanity.

Towel holders

The towel holders in most bathrooms are bars, but rings also account for a lot of towel hanging. Bars are probably more practical (Fig. 15-3), but the ring is often considered more decorative. I usually give my customers both. In addition to bars and rings (Fig. 15-4), you have the option of offering a holder that will hold multiple towels in a folded condition. These devices are similar to those found in many motels. Multitowel holders are not as attractive, usually, as some other types of holders, but they are very practical in terms of large families who use a common bathroom. Since towel holders are a typical accessory that you will probably be expected to install, it would be a good idea

Figure 15-3

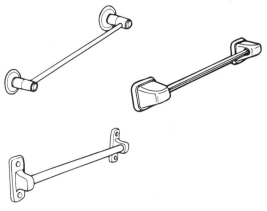

Towel bars. Courtesy of Eljer

Figure 15-4

Towel rings. Courtesy of Eljer

to gather a good selection of material descriptions and photos for your customers to choose from.

Soap dishes

Soap dishes are not always installed in modern bathrooms, but plenty of people do still like these accessories. If you are having ceramic tile installed around a tub or shower, the soap dish for this fixture will be built as a part of the tile wall. Molded bathing units will have their soap dishes already precast into the body of the unit. As a contractor,

Figure 15-5

Soap dishes. Courtesy of Eljer

your work with soap dishes will most likely be around the lavatories in bathrooms. There are, once again, plenty of styles (Fig. 15-5) and types to choose from. I like the wooden soap dishes that have plastic inserts in them. Homeowners can remove the plastic inserts and clean them easily. Some customers will not want soap dishes around their lavatories at all, feeling that they are disgusting to look at. Also, the increased use of liquid soap has eliminated some of the use of soap dishes. All you have to do is look around a bathroom supplier or catalog to find a myriad of possibilities for soap dishes.

Toothbrush holders

Toothbrush holders seem to remain popular with homeowners. Many of these holders incorporate a cup holder into the center of the rack. The problem with toothbrush holders is that they tend to get dirty quickly. Toothpaste dripping from brushes that are not rinsed thoroughly builds up on the holders. For this reason, some homeowners request that the holders be installed on the interior of vanity doors, which really isn't a bad idea. The holders may be made of a glasslike finish, wood, chrome, or some other designer finish. Any way you cut it, you will probably be installing some form of toothbrush holder in most bathroom jobs.

Cup holders

Cup holders (Fig. 15-6) don't seem to be very popular in the areas where I work. However, there are accessories available for holding cups, and this might be another request that you will get from your customers. Based on personal experience, I don't think that cup holders will be found on many jobs, but perhaps you should put some literature on them in your files for the customers who are interested in such items. As with most other bath accessories, installation of cup holders is not a problem, once you know where your customer wants the device placed.

Figure 15-6

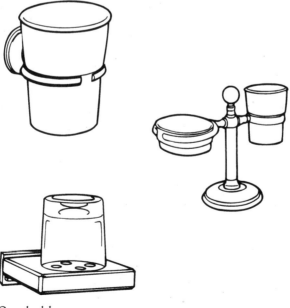

Cup holder. Courtesy of Eljer

Mirrors

Mirrors are probably one of the most sought-after bathroom accessories on the market. They can also be a very expensive option to purchase. When you sell a job, you might be thinking about a simple mirror that is flat glass and that will attach to the wall over the lavatory with plastic clips. Your customers may have a very different type of mirror in mind, like one that cost a few hundred dollars more! If you will be supplying mirrors for your jobs, make sure that you clarify the costs with your customers. One approach that works well is the same approach that many contractors use for lighting fixtures; give your customers a mirror allowance to work with. However you do it, define clearly how expensive a mirror, or mirrors, can be without exceeding the price that you bid the job at.

Mirrors are important in bathrooms, and they can have a lot to do with the overall appearance of the room. Flat edges, beveled edges, square mirrors, rectangular mirrors, edged mirrors, and a full array of other mirrors all combine to confuse customers. Believe me when I say that mirrors can cost hundreds of dollars. If you figure a job with a $25 mirror budget, without having your customer agree to the committed price, you could find yourself out of some serious pocket change before you can satisfy your customer. Are you going to provide a full-length mirror in the bathroom? Will the mirror over the lavatory be the only one in the room? Does your customer expect a huge mirror for the makeup portion of the extended vanity cabinet? It doesn't take long to get into deep financial ruts with the prices of large or fancy mirrors.

Hardware

You might not think much about hardware when you think about bathrooms. This can be a mistake. There are a number of companies who specialize in nothing but hardware. The hardware (Figs. 15-7 & 15-8) can be used on doors, windows, vanities, and so forth. With the right combination of hardware, you can create a bathroom that is

Figure 15-7

Cabinet hardware. Courtesy of Horton Brass

close to unique, and this is good for your image as a builder or remodeler. There is no shortage of hardware available to you. Plenty of companies and suppliers will provide you with color catalogs; all you have to do is ask for them. A high-quality cabinet will come with good hardware, but if you replace the factory issue with custom hardware, people will notice the customized touch. In the scheme of things, hardware is inexpensive for the distinction that it gives a bathroom. Cabinet pulls (Fig. 15-9) and hinges will normally comprise most of the hardware that you will put into a bathroom, but the door set is another piece of hardware where you can get creative.

Ceiling hooks

Ceiling hooks may seem like an odd accessory for a bathroom, but many people like to have hanging plants in there bathrooms. This is especially true of large bathrooms. If your customers will want ceiling hooks installed, you or your crew should contract to install the hooks for them. It's not that you will make much more money in doing this, but you will have the ability to preserve the quality of the ceiling you just created. Homeowners often fail to hit joists with their ceiling hooks. It would be a shame to have a new ceiling full of false holes. Again, this is the type of accessory few people think of, but it is one that a lot of people add after a room is done. If you have the chance, install the hooks for your customers to maintain a professional look to your ceiling.

Figure 15-8

Door latches. Courtesy of Horton Brass

Figure 15-9

**BURNISH
BRASS PULL**

C2035

3" centers.

**POLISHED
BRASS KNOB**

C2053

1¾₆"diameter.

**CHROME
PULL**

C2037

3" centers.

**CHROME
KNOB**

C2049

1¾₆" diameter.

Cabinet pulls. Courtesy Wellborn
Cabinets

**COLONIAL
BRASS PULL**

C2039

3" centers.

**STIRRUP
BRASS PULL**

C2041

3" centers.

**BRASS PULL
W/WHITE
CERAMIC
INSERT**

C2043

3" centers.

On and on

We could go on and on with bathroom accessories. Baskets for vanities, waste baskets, and a host of other decorative items could be discussed. Borders and stenciling are common accessories that are added in afterthought. Spend time talking with your customers to see what types of accessories they will want. Strive to get as many of them as possible on your contract before a job is started. Use change-order forms to document additional work as it develops, and you can probably count on new items coming up as a job progresses. Few people can think of everything that they want all at once. Most importantly, cover all the bases before you pull off of a job. Try to have competent professionals install as many of the accessories as possible. When you leave a job, you want it to be as perfect as possible. After all, your company name is on the line. The ultimate key to success in bathroom work is good communication with your customers. If you can accomplish this, you should enjoy a long and prosperous career in the field of bathroom construction and remodeling. I wish you the best in all your business endeavors.

Glossary

ABS PIPE A type of plastic pipe frequently used in plumbing. The letters, ABS, are an abbreviation for Acrylonitrile Butadiene Styrene. It is black in color and is most frequently used in the form of schedule 40 pipe.

ADJUSTABLE RATE LOAN A loan with a flexible interest rate, tied to a common reference index. This index can be a Treasury Bill Rate, the Federal Reserve Discount Rate, or some other agreed upon index. These loans typically start at a lower interest rate and escalate over a period of five years. After five years, most of these loans stay at a fixed rate for the remainder of the loan.

AIR GAP The most common use of this term in remodeling is in reference to the device a dishwasher drain is connected to. These devices protect a dishwasher from the risk of back siphoning contaminated water from the drainage system into the dishwasher.

ANNUAL APPRECIATION RATE The rate at which real estate values increase each year.

APPRAISER An individual used to determine the value of real estate. Appraisers may be certified or uncertified; in many states, appraisers are not required to pass stringent licensing requirements.

ARCHITECTURAL PLANS Blueprints designed and drawn by an architect.

BAIT AND SWITCH SCAM A frequently used unethical marketing ploy to make people respond to advertising. The advertising will entice an individual to come to the advertiser, but the advertised price or product will not be available. The advertiser will attempt to sell the consumer a different product or a similar product at a higher price.

BALLOON PAYMENT A single lump-sum payment, frequently associated with real estate loans. These payments are generally due and payable in full after an agreed upon time.

BAND BOARD The piece of lumber running the perimeter of a building and attached to the floor joists. Normally a wooden member of a size equal to the attached floor joists and placed on the outside wall end of the floor joists. A band board provides a common place for all floor joists to attach and maintains stability and proper alignment.

BASEBOARD TRIM A decorative trim placed around the perimeter of interior partitions. Used where the floor covering meets the wall to create a finished and attractive appearance.

BID PHASES The different phases or aspects of work to be performed and priced. Examples would include plumbing, heating, electrical, roofing, and all other individual forms of remodeling or construction.

BIDS Prices given by contractors and suppliers for labor and material to be supplied for a job.

BLOCK NUMBER One of the elements needed for a legal description of real property in areas segregated into blocks.

BLUEPRINTS The common name of working plans, a type of plan printed in blue ink and showing all aspects of the construction methods to be used in building and remodeling.

BOILER A type of heating system, usually designed to provide hot water heat from baseboard radiation.

BOW WINDOW A window projecting outward, beyond the siding of a home and supported by its own foundation or support beams. Sometimes referred to as a bay window.

BREEZEWAY A covered and sometimes enclosed walkway from one point to another. Commonly used to connect a garage to a house when direct connection isn't feasible or desirable.

BTU British Thermal Unit; an industry standard in the measurement of the amount of heat needed for an area. One BTU equals the amount of heat required to raise a single pound of water one degree in temperature.

BUILDER GRADE A trade term meaning a product of average quality and normally found in production built housing.

CALL-BACK A trade term referring to a warranty service call. A form of service call where the contractor is not paid for the services rendered, due to the nature of the problem.

CANTILEVER Refers to a building practice where the wooden frame structure extends beyond the foundation. Canterlevers are created when the floor joists overhang the foundation.

CARDBOARD STYLE CABINETS A trade term referring to inexpensive production cabinets of low quality.

CARPET PAD The support, generally foam, between the carpet and subfloor or underlayment.

CASEMENT WINDOW A window with hinges on the outside and a mechanical crank to open and close the window. These windows open outwardly and are typically very energy efficient.

CAULKING A compound used to fill cracks and provide a satisfactory finished service.

CEILING JOIST Structural members providing support for a second story floor and a nailing surface for a lower story's ceiling.

CERAMIC TILE A product used for floors, countertops, wall coverings, and tub or shower surrounds. The most common

▲

ceramic tiles are approximately four inches square and are comprised of a pottery type of material.

CERTIFICATE OF OCCUPANCY These are issued by the local codes enforcement office when all building code requirements are met. They allow the legal inhabitance of a dwelling or business property.

CHAIR RAIL A wooden member, of finished trim quality, placed horizontally at a point along the walls where chairs would be likely to come into contact with the wall. Chair rail serves some practical purpose, but is most frequently used as a decorative trim in formal dining rooms.

CHANGE ORDER A term applied to a written agreement allowing a change from previously agreed to plans. Change orders detail the nature of the change and all pertinent facts affected by the change.

CHINA LAVATORY A bathroom wash basin made of vitreous china. These lavatories provide a clean, nonporous service for a wash basin and usually have a shiny finish.

CHRONOLOGICAL ORDER The order in which events should happen, such as, in planning a schedule of events or work to be performed.

CIRCUIT BREAKER The modern equivalent to the old style electrical fuses. These devices add protection from overloaded electrical circuits by shutting down the circuit if it is producing a dangerous electrical current.

CLOSE COUPLED FAUCET Also referred to as a four-inch center faucet, these faucets are produced as an integral, one piece unit. The handles and the spout are molded from the same material, producing a faucet with all working parts molded together.

CLOSED SALE A closed sale is a consummated or settled sale. It is a completed transfer of the property being sold.

CLOSING COSTS These are expenses incurred to settle a loan transaction. They can include legal fees, appraisal fees, survey fees, insurance, and other related expenses.

CLOUDED TITLE When a title is clouded, there are unanswered questions or liens about or against the title. These conditions can make the transfer of the property to another owner very difficult.

CODE ENFORCEMENT OFFICER An authorized representative of the building code enforcement office. The individual responsible for the approval or denial of code inspections and the party responsible for issuing a certificate of occupancy.

COMMERCIAL GRADE CARPET Normally a close weaved and very durable carpet, suitable for heavy traffic and abuse. This carpet is designed for easy cleaning and to handle the most demanding traffic without undue wear.

COMPARABLE SALES SHEET A form used to compile information on real estate activity in an area allowing an accurate appraisal of a property.

COMPARABLE SALES BOOK Generally produced by Multiple Listing Services, these books reflect a history of all closed sales for the last quarter of a given year. These books are used to determine appraised values of the subject properties.

COMPETITIVE GRADE FIXTURE A trade term referring to an inexpensive fixture, normally found in tract housing or starter homes.

COMPLETION CERTIFICATE A document signed by the customer acknowledging all work is complete and satisfactory.

COMPRESSION FITTING A type of fitting used to make a plumbing connection. Typically utilizes a brass body and nut with a ferrule to compress over the pipe, preventing water from leaking.

CONCRETE APRON The section of concrete where a garage floor joins the driveway. Aprons allow for a smooth transition from a lower driveway to an elevated garage floor.

CONTRACT DEPOSIT A financial deposit given when a contract is signed and before work is started.

CORNICE A horizontal molding usually projecting from the top of an exterior wall to provide better water drainage.

COSMETIC IMPROVEMENTS An improvement with no structural significance, performed as an aesthetic enhancement to the property.

COST INCREASE CAP An amount of money set to limit the price increase of a future purchase.

COST APPROACH An appraisal technique used to determine a property value, based on the cost to build the structure.

COST ESTIMATE SHEET A form designed to accurately project the cost of a proposed improvement.

CRAFTSMEN A word used to describe a person, working in a trade, who is experienced and proficient in the trade.

CRAWL SPACE The space beneath a house, between the first story floor joists and the ground, surrounded by a foundation.

CROWN MOLDING A decorative wood trim placed at the top of an interior wall, where the wall meets the ceiling.

CURB APPEAL A term used in real estate sales referring to the exterior appearance of a property.

CUT SHEETS Illustrated fact sheets providing detailed information on a product.

DAYLIGHT BASEMENT A basement with windows allowing natural light to flow into the basement.

DECKING Decking can apply to the materials used to build and exterior deck or the material used to build interior flooring systems.

DEED DESCRIPTION A legal description of a property as it is referred to in the registered deed to the property.

DEMO WORK Demolition work, the process of dismantling or destroying existing conditions.

DEMOGRAPHIC STUDIES Statistical studies of the population. These studies can include specifics such as age, sex, income, and other highly detailed information.

DENIAL NOTICE A notice of rejection or turn-down for a requested service, such as a loan.

DIRECT MARKET EVALUATION APPROACH An appraisal technique used to determine property value by comparing the subject property to other similar properties. All pertinent features of the subject property are compared to similar properties and financial adjustments are made for differences to establish a value on the subject property.

DISCRETIONARY INCOME Income not committed to a particular expense. The amount of money an individual has to spend on anything they wish.

DORMER A projection built from the slope of a roof allowing additional room height and the opportunity to install windows.

DOWNTIME A period of nonproductive or lost time.

DRAIN A pipe carrying water or waterborne waste to a main drainage system.

DRYWALL MUD Joint compound, the substance used to hide seams and nail or screw heads in the finished walls of a home.

DRYWALL A term used to describe a type of wall covering made of gypsum.

DWV SYSTEM Drain, Waste, and Vent system, the plumbing system used in a home for the drainage and venting of plumbing fixtures.

EIGHT-INCH-CENTER FAUCET A faucet designed to have the two handles spaced on an eight-inch center. These faucets are comprised of separate elements for each handle and the spout; they are connected beneath the fixture's surface to allow the faucet to operate.

EIGHTEEN-INCH TOILET A special toilet designed for the physically restricted person. The seat of these toilets is higher than a standard toilet, allowing easier use without as much demand for physical strength.

ELECTRIC HEAT A trade term referring to electric baseboard heating units attached permanently to the interior wall of a home. Electric heat can utilize other forms of heating equipment, such as a wall mounted blower unit. Electrical current provides the source for producing heat from this type of heat.

ELECTRICAL SERVICE A trade term referring to the size and capacity of a home circuit breaker or fuse box. Older homes were equipped with 60 AMP electrical services. Modern homes have 100 AMP or 200 AMP electrical services as a standard service.

ELEVATIONS A term used with drafting and blueprints referring to illustrations on the blueprints. Examples are picture drawings of the front of a house, the side of the house, and the rear of the house.

ESTIMATED JOB COST The projected cost required to complete a job.

EXCLUSIONS Phrases or sentences releasing a party from responsibility for certain acts or circumstances.

EXTERIOR WALL SHEATHING The exterior wall covering placed between the exterior wall studs and the exterior siding.

EXTERIOR IMPROVEMENTS Improvements made outside of a dwelling. Examples are garages, landscaping, exterior painting, and roofing.

EXTERIOR FRAMING The material or labor used in the construction of exterior walls and roof structures.

FAIR MARKET VALUE The estimated value of a property to the buying public in the real estate market.

FIBERBOARD A composite sheet made from pressed materials bonded together for use as a wall sheathing.

FILL-IN-JOB A trade term for a job with no committed completion time. These jobs are often done at discounted prices because they allow flexibility for the contractor.

FINISHED BASEMENT A basement that has been completed into finished living space. The walls, ceiling, and floor are all completed to an acceptable finished standard. The basement is provided with heat, electrical outlets, lights, and switches.

FIXED RATE LOAN A loan with a fixed interest rate. The interest rate does not fluctuate, it remains constant for the life of the loan.

FLASHED A trade term applying to the attachment of articles to houses or roofs and the penetration of roofs by pipes. When these conditions exist, they are flashed to seal the area from water infiltration. Plumbing pipes exiting through a roof are flashed with neoprene, or some other material to prevent water leaks around the pipes. Where decks or bow windows are attached to a house, they are flashed with lightweight metal to prevent water damage behind the point of attachment.

FLOOR JOIST A structural member or board used to support the floor of a house. Floor joists span between foundation walls and girders at regular intervals to provide strength and support to the finished floor.

FLUORESCENT LIGHT A lamp or tube producing light by radiant energy, a tube coated with a fluorescent substance giving off light when mercury vapor comes into contact with electrons.

FOOTING A support, usually concrete, under a foundation providing a larger base than the foundation to distribute weight. Footings are placed on solid surfaces and reduce settling and shifting of foundations.

FORCED HOT AIR FURNACE A type of heating system producing warm air heat and forcing the warm air through ducts, with the use of a blower, into the heated area.

FORM CONTRACT Standard or generic forms, available at office supply stores and stationery specialty vendors, intended for use as legal contracts when the blank spaces are filled in.

FOUNDATION The base of a structure used to support the entire structure.

FOUR-INCH-CENTER FAUCET The same as a close coupled faucet. These faucets are produced as an integral, one piece unit. The handles and the spout are molded from the same material, producing a faucet with all working parts molded together.

FRAMING A trade term referring to the process of building the frame structure of a home for siding, sheathing, and wall coverings to be applied to.

FUMIGATION The process of exposing an area to fumes to rid the area of existing vermin or insects.

FUNCTIONAL OBSOLESCENCE An appraisal term referring to the absence of common desirable features in the design, layout, or construction of a home. A kitchen without cabinets or a modern sink would be a form of functional obsolescence.

GENERAL CONTRACTOR The contractor responsible for the entire job and the person who coordinates subcontractors in individual aspects of the job.

GRACE PERIOD A term referring to the period of time a commitment may be unkept before enforcement action is taken.

GREENHOUSE STYLE WINDOW A bow window unit designed to extend beyond the exterior wall of a house and made mostly of glass, including the roof portion of the window. The window is meant to allow additional lighting and provide a feeling of openness.

GROUND FAULT INTERCEPTOR OUTLET An electrical outlet, used primarily in bathrooms, with a safety feature to protect against electrical shock.

GROUT The substance used to fill cracks between tile during the installation process.

HARD COSTS Expenses easily identified and directly related to a job. Examples include the cost of labor and materials for the construction process.

HEAT PUMP A device used for heating and cooling a home. Heat pumps are an effective heating and cooling unit in moderate climates and are being improved for use in extremely cold climates. They are energy efficient and do not require a flue or a chimney.

HEAT LAMP, FAN COMBINATION An electrical fixture commonly placed in the ceiling of bathrooms. The unit combines an exhaust fan with a heat lamp providing warmth and the removal of moisture.

HVAC Heating, Ventilation, and Air Conditioning.

IN THE FIELD A trade term referring to being out of the office and on the job.

INCANDESCENT LIGHT A light using a filament contained in a vacuum to produce light when heated by electrical current.

INSULATED FOAM SHEATHING A type of sheathing made from compressed foam and covered by a foil or other substance allowing its use as a wall sheathing with increased insulating value.

INTERIOR PARTITIONS The walls located within a home dividing the living area into different sections or rooms.

INTERIOR REMODELING Altering the condition of areas within the home.

INTERIOR TRIM A trade term broadly referring to any decorative wood trim used within the home. Examples are baseboard trim, window casing, chair rail, crown molding, and door casings.

JOB JUGGLING A trade term referring to the practice of moving among multiple jobs during the same time frame. This practice is frequently used to describe ineffective work habits performed when a company has taken on too much work or has a slow cash flow.

JOINT COMPOUND Also know as drywall mud, it is the substance used to hide seams and nail or screw heads in the finished walls of a home.

JOISTS Supporting structural members, usually made of wood, allowing the support of floors and ceilings.

LIABILITY INSURANCE Insurance obtained to protect the insured against damage or injury claims and lawsuits.

LIEN WAIVER A document used to protect property from mechanic and materialman liens. These documents are signed by the vendor upon payment to acknowledge the payment and to release their lien right against the property the products or services were rendered for.

LIEN RIGHTS The right of contractors and suppliers to lien a property where services or products are provided, but not paid for.

LINE DRAWING A simple plan drawn with single lines indicating the area's perimeters and division into sections and rooms.

LINEAR FEET A term used to describe a unit of measure, measuring the distance between two points in a straight line.

LOAD BEARING WALLS Walls supporting the structural members of a building.

LOAN COMPANY Companies specializing in making loans, but not offering the services of a full-service bank or savings and loan association.

LOT NUMBER A number assigned to a particular piece of property on zoning or subdivision maps.

LOWBOY TOILET A one-piece toilet with a low profile. These toilets have an integrated tank and the tank does not rise as high above the bowl as a standard toilet.

MARKET ANALYSIS A study of real estate market conditions used to establish an estimated fair market value for the sale of a home.

MARKET EVALUATION A term used interchangeably with market analysis, a study of real estate market conditions used to establish an estimated fair market value for the sale of a home.

MARKETABILITY A term used to describe the feasibility of selling a house on the current real estate market. Marketability is determined by the features and benefits of a home.

MATERIAL A trade term referring to the products and goods used in building and remodeling.

MATERIAL LEGEND An area on blueprints describing in great detail the type of materials to be used in the construction or remodeling of the proposed project.

MATERIALMAN'S LIEN A recorded security instrument placed on the title of a property to secure an interest in the property until a legal dispute can be resolved. These liens are placed when a supplier has supplied material for a property and has been refused full payment. Liens create a cloud on the title and make transferring ownership of the property very difficult.

MECHANIC'S LIEN A recorded security instrument placed on the title of a property to secure an interest in the property until a legal dispute can be resolved. These liens are placed when a contractor has supplied labor for a property and has been refused full payment. Liens create a cloud on the title and make transferring ownership of the property very difficult.

METES AND BOUNDS The oldest method of describing the boundaries of a property. In this method the boundaries are described in detail using natural or artificial monuments and by explaining the direction and distance the property lines run.

MITER BOX A small box with no top and slits placed in the top of each side to allow a saw blade to pass through the box and cut wood laid in the box. These slits are cut to guide the saw along a specific angle when the wood is cut. They are used for cutting angles on finish trim boards.

MORTAR A mixture of lime or cement used between bricks, blocks, and stones to hold them in place.

MORTGAGE BROKER A liaison between a borrower and a lender. An individual who attempts to procure financing for a client for a fee.

MORTGAGE A pledge or document securing a lender's investment and accompanying the note for the loan.

MULTIPLE LISTING SERVICE A service provided to real estate brokers combining a listing of all real estate for sale by the members of the service. Multiple listing books provide information on all homes for sale and sold during a given time by the participating members in the real estate profession.

NONSTRUCTURAL CHANGES Changes not affecting the structural integrating of building. Examples are replacing kitchen cabinets, installing new carpet, and painting.

NONCONFORMING A house or improvement not being similar to surrounding properties in age, size, use, or style. An example would be a one-level Ranch style house in a neighborhood comprised of two-story Colonial style homes.

OAK VENEER VANITY A vanity constructed of plywood or particleboard and covered with an exterior finish of oak.

ONE-PIECE TOILET A modern style toilet with the tank and the bowl molded as a single element, creating a sleek appearance and an easy-to-clean surface. Standard toilets have the tank and the bowl, as two separate pieces, joined together with brass bolts and nuts.

OPEN-END BILLING A trade term referring to working on a time and material basis and billing for all labor and material involved in a job, with no limit on the total amount to be billed.

OUTLET PLATE A trade term describing the cover placed over an electrical outlet and screwed to the center of the outlet.

OUTSIDE WALL Any wall with one side meeting outside air space.

OVERBUILDING A term describing the practice of investing money in a home that is unlikely to be recovered, due to surrounding properties. An example would be adding three bedrooms to a home, for a total of six bedrooms, when surrounding houses only have three bedrooms.

PARTICLEBOARD A composite of wood chips bonded and pressed together to create a sheet to be used for subflooring or sheathing.

PEDESTAL SINK A prestigious bathroom sink with a china bowl hung on the wall and supported by a china pedestal. The pedestal adds support to the bowl and hides the plumbing connected to the sink and faucet.

PERMITS Documents issued by the code enforcement office allowing work to be legally performed.

POLYBUTYLENE PIPE A modern type of flexible plastic pipe used for the distribution of potable water in building.

PLUMBING STACK A pipe rising vertically through a building to carry waste and water to the building sewer or to vent plumbing fixtures when it penetrates the roof of the building.

PLYWOOD A wood product comprised of multiple layers of veneer joined with an adhesive. Plywood usually has three layers, but can have more and it is always comprised of wood veneers in odd numbers. Typically, the grain of each veneer is joined at ninety degree angles.

POINT-UP The procedure used to repair or replenish the mortar between bricks, blocks, stone, and tile.

POINTS Also known as discount points, they are fees paid to a lender to increase the yield of a loan being offered by the lender.

POTABLE WATER Water meeting the requirements to be considered safe for drinking, cooking, and domestic purposes.

POWDER ROOM A trade term referring to a room containing a toilet and a lavatory, without containing a bathtub or shower.

PREFAB TRUSS SYSTEM A manufactured roof system eliminating the need to stick build a rafter roof. Trusses only need to be set into place and secured, they require no on-site cutting or building.

PREPAYMENT PENALTY A penalty charged by a lender when a loan is paid in full before its maturity date. Prepayment penalties insure the lender of receiving the full yield of a loan, regardless of when it is paid off.

PRESSURE BALANCE CONTROL A trade term used to describe a type of plumbing faucet. These faucets are considered a safety feature because they prevent the user from being scalded by hot water if there is a fluctuation in the cold water pressure.

PREVENTIVE IMPROVEMENT Improvements designed to reduce costly repairs and replacements through routine maintenance.

PRIME FOR PAINT The process of preparing a surface to receive paint. This procedure produces better results than when paint is applied without a primer.

PRODUCTION SCHEDULE The agenda for events to be performed in the construction and remodeling process.

PROGRESS PAYMENT Periodic payments made as work progresses into defined stages, such as rough-in and final.

PUNCH-OUT A trade term referring to the process of correcting deficiencies and making minor adjustments at the end of the job.

PVC PIPE Poly Vinyl Chloride, a type of plastic pipe used in plumbing. Frequently used for drains and vents and occasionally used for cold water piping.

QUOTES Firm prices given by contractors and suppliers for labor and material.

RAFTER CUTS A trade term for the angles cut on rafter boards when stick building a roofing system.

RAFTERS Structural members, usually made of wood, supporting the roof of a building.

REGISTRY OF DEEDS A place where deeds are recorded and available for public inspection.

REHAB Reconstruction or restoration of an existing run-down building.

REMODELING The practice of altering existing conditions and adding new space to existing structures.

RETAINAGE A holdback of money owed to a contractor for an agreed upon period of time, to protect the consumer from defective material or workmanship.

RIGID COPPER TUBING Frequently called copper pipe, rigid copper tubing is a common material used in the potable water distribution system of residences. It typically comes in rigid lengths, ten or twenty feet long, and can be cut as needed.

RIMMED LAVATORY A drop-in style lavatory with a steel rim surrounding the lavatory bowl to hold it in place.

RIP-OUT A trade term referring to the removal of existing items to allow the installation of new items. An example would be removing an old bathtub and surrounding tile, to allow for the installation of a new tub and surround.

ROOF SHEATHING The material secured to the rafters or trusses to allow the installation of a finished roof. Plywood and particleboard are frequently used as roof sheathing.

ROOFING FELT A black paperlike product applied between the roof sheathing and the shingles. It reduces the effects of extreme temperature and moisture.

ROUGH PLUMBING This term refers to the pipes and fittings of a plumbing system, but does not include fixtures.

ROUGH-IN DRAW A progress payment made when the rough-in work is complete.

ROUGH-IN A trade term referring to the installation of material prior to enclosing the stud walls. Examples would be for plumbing, heating, and electrical systems. The bulk of these systems must be installed before the wall coverings are applied, this is considered rough-in work.

ROUND FRONT TOILET A toilet with a rounded bowl, as opposed to a toilet with an elongated bowl. Most residential toilets have round fronts.

PLANS SCALE A defined and constant unit of measurement for blueprints and drawings. An example is standard blueprints utilize a scale where each quarter of an inch on the blueprints equals one foot in the actual building.

SCHEDULE 40 PIPE This is a rating for the thickness and strength of a pipe; it is the standard weight of plastic pipe used for residential drainage and vent plumbing systems.

SEAL FOR PAINT When stains exist and must be painted over, you should apply a sealing agent to the stain before painting. This process prevents the stain from bleeding through the new paint.

SECONDARY MARKET A financial term referring to markets where banks and other lenders sell their real estate loans. The original lenders normally continue to service these loans for a fee, but are able to recycle their available lending funds by selling the loans to the secondary market. The secondary market is comprised of individual investors, corporations, and organizations.

SECTION OF PROPERTY A term used in providing a legal description of a property. The section of a property is typically referred to on zoning and subdivision maps.

SELLING-UP A trade term pertaining to selling a customer additional services and products. The act of enticing a customer to spend additional money, above and beyond the original contract amount.

SETBACK REQUIREMENT A term relating to zoning regulations, where a certain amount of unobstructed space must exist between properties. With setback requirements, owning the land does not mean you can build on all of it. These requirements establish a rule on how far a structure must be from each property line.

SHEATHING The material applied to exterior studs and rafters to allow the installation of finished siding and roofing.

SHEET VINYL FLOORING Also known as resilient sheet goods, these floor coverings are available in widths of six, nine, and twelve feet. They are a common and well-accepted floor covering for kitchens and bathrooms.

SHIMS Small pieces of tapered wood used to level construction and remodeling materials, such as, doors, cabinets, and windows.

SHOWER HEAD ARM OUTLET The female adapter located approximately six feet and six inches above the finished bathroom floor, in the center of the shower area. The location where a shower head arm is screwed into the adapter, allowing the use of the shower head.

SILL The board placed on top of the foundation and beneath the floor joists.

SITE WORK Normally includes excavation, but always refers to the preparation of a site for construction.

SITE CONDITIONS A term used when describing the conditions of a construction site. Examples would be level, sloping, rocky, and wet.

SKYLIGHT A glass panel located in the roof, allowing natural light to fill a space below it.

SLAB Usually a flat interior reinforced concrete floor area.

SOFT COSTS Expenses incurred in a project that are not directly related to construction or remodeling in the strictest sense. Examples are loan fees, surveys, legal fees, and professional fees.

SPA A bathing tub with whirlpool jets designed to hold and heat water indefinitely with the use of chemicals and an independent water heater. Spas are designed to be filled and drained with a garden hose.

SPECIFICATIONS This term applies to a compilation of the services and products to be used in the completion of a project. Specifications can be addressed individually in a contract or may take the form of a separate collection of documents.

SQUARE FEET This term is a unit of measure frequently used by contractors. To obtain the square footage of an area, you must multiply the length of two perpendicular walls together. In a rectangular space, this procedure will give you the total square footage of the area.

SQUARE FOOT METHOD An appraisal technique where a value is assigned for each square foot of space contained in a building. This method is reasonably accurate with standard new construction procedures, but is rarely accurate or used in remodeling.

SQUARE YARDS This term is a unit of measure most commonly used in floor coverings. To obtain square yardage you must take the square footage of an area and divide it by nine.

STANDARD GRADE FIXTURE This is a trade term used interchangeably with a builder grade fixture. A product of average quality and normally found in production built housing.

STICK BUILD A trade term meaning to build a structure on-site with conventional construction methods.

STORAGE CEILING JOISTS Ceiling joists rated to carry an additional weight load for storage above the ceiling.

STRIP LIGHTS Multiple incandescent lights mounted on a metal or wood strip. Common applications include three or four lights mounted on an oak strip for use in bathrooms.

STRUCTURAL WORK Work involving the structural integrity of a building. Examples are adding a Dormer addition, expanding an existing structure, or relocating load-bearing walls.

STRUCTURAL MEMBERS Structural members normally consist of wood and they support a portion of the building. Examples are floor joists, rafters, and ceiling joists.

STRUCTURAL PLANS Plans or blueprints detailing the materials to be used and the placement of these materials for structural additions or changes.

STRUCTURAL INTEGRITY The strength of a structure to remain in its planned position without fail.

STUDS Typically made from wood in residential applications, studs are the vertical wooden members of a wall. They are placed at regular intervals to allow support and a nailing surface for wall coverings and exterior siding.

SUBFLOOR Generally either plywood or particleboard sheets attached to floor joists under the finished floor covering.

SUBCONTRACTOR A contractor working for a general contractor. Examples could be plumbers, electricians, or HVAC contractors.

SUBS A trade term abbreviation for subcontractors.

SUPPLIERS A trade term referring to the companies supplying materials used in construction and remodeling.

SUPPORT COLUMNS Vertical columns used for structural support. An example could be the columns found in basements or garages, supporting the main girder.

SUBSTITUTION CLAUSE A common clause in contracts allowing a supplier or contractor to substitute a similar product in place of the specified product. These clauses should only be allowed when they are heavily detailed and clearly define the products substitutions will be made with.

T & M A trade term meaning Time and Material. A form of billing for all labor and material supplied with no limit to the billed amount.

TAKE-OFF A trade term meaning an estimate of the materials and labor required to do a job. Take-offs are generally associated more with material than labor.

TANKLESS COIL An internal part of a hot water boiler heating system, also referred to as a domestic coil. The coil provides a source of potable hot water by heating water passing through a copper coil found beneath the boiler's jacket.

TAPED DRYWALL A trade term denoting drywall that has been hung and taped. The tape is applied with the use of joint compound to hide the seams where sheets of drywall meet.

TEMPERATURE CONTROLLED FOUNDATION VENTS
Modern foundation vents able to sense temperature and open or close automatically. These vents allow for better foundation ventilation throughout the year.

TEMPLATES A trade term with multiple definitions. The first definition is a plastic stencil kit allowing draftsmen to draw consistent symbols of items for blueprints. Examples are toilets, doors, electrical switches, and sinks. The second use of the word is to describe a guide. These guides are used for cutting countertops to allow the installation of kitchen sinks and other related work.

TENTED FUMIGATION The process of enveloping an entire house in a tent to allow a total fumigation of all the home's wood related products. This process is used to remove certain wood investing insects.

THRESHOLD A trim piece which connects two flooring areas. Usually made of wood, metal, vinyl, or marble. These usually serve to trim the seam between two different materials, such as a vinyl bathroom floor and the hall carpet.

TIME AND MATERIAL BASIS Basically the same as open-billing; a form of billing for all labor and material supplied with no limit to the billed amount.

TITLE SEARCH A function frequently preformed by attorneys to certify a title is clear of liens or other clouds preventing a satisfactory real estate closing.

TOILET BOWL The part of a toilet where the seat is attached.

TOILET TANK The part of a toilet where the handle is for flushing a toilet. A reservoir tank holding up to five gallons of water to allow the toilet to be flushed.

TONGUE AND GROOVE PANELING A type of paneling or siding with a groove on one side and a projection on the other. The projection is placed inside the groove of adjoining panels to form an attractive finished seam.

TRACT HOUSING A trade term describing production or subdivision housing. The term refers to houses built on a tract of land.

TRADESPEOPLE People working within a trade.

TRIGGER POINTS A sales term referring to the key words or actions resulting in a potential buyer agreeing to make a purchase.

TRIM A trade term applied to any object, normally wood, used to provide a finished look to a product or installation.

TWELVE-INCH-ROUGH TOILET A standard toilet where the center of the drain pipe is located twelve inches from the finished wall behind the toilet.

TWO-PIECE TOILET A combination of a toilet tank and bowl, connected by brass bolts and nuts, to form an operational toilet.

TYPE "M" COPPER TUBING The type of copper refers to the thickness of the wall of the tubing. Type "M" copper is marked with a red stripe and is frequently used for piping residential hot water heating systems. Until recently, it was a common carrier of potable water to plumbing fixtures. It is not approved in most locations for underground use; code revisions are phasing out type "M" as a potable water distribution pipe and requiring type "L."

TYPE "L" COPPER TUBING The type of copper refers to the thickness of the wall of the tubing. Type "L" copper is marked with a blue stripe and is approved for use underground and has a thicker sidewall than type "M" copper. It is becoming the most frequently used copper in residential water distribution systems.

TYPE "K" COPPER TUBING Type "K" copper is marked with a green stripe and has a thicker wall than type "M" or type "L" copper. It is the preferred choice of copper tubing for underground installations, but sees little use in above ground, residential applications.

UNDERLAYMENT A trade term for a smooth sheet of wood applied between the subfloor and the finished floor. Underlayment is most commonly used when vinyl sheet goods are to be installed. It provides a smooth, even surface for the vinyl to rest on.

UNFINISHED BASEMENT A trade term describing a basement with a concrete floor and unfinished walls and ceiling. There are minimal electrical outlets and little to no heat in an unfinished basement.

VALANCE A short curtain forming a border between a window and the ceiling or a short trim board connecting the top of kitchen cabinets to the ceiling.

VANITY A trade term describing a base cabinet for a bathroom lavatory or sink.

VENT PIPE A part of the plumbing system designed to allow the free circulation of air within the plumbing drain and vent system.

VENTED EXHAUST FAN The two most common locations for vented exhaust fans are the kitchen and the bathroom. In the kitchen, the fan vents the fumes and gases from the range to the outside air. In a bathroom the exhaust fan is designed to remove odors and moisture from the room. Both types of fans are vented to the outside air space with a duct or coiled hose.

VINYL SIDING A type of exterior siding requiring little to no maintenance with a life expectancy of twenty years. The color is a part of the molded vinyl and will not fade or wear off under normal conditions.

WAINSCOTTING The procedure of installing wood on the lower portion of a wall, joined by chair rail, to meet the upper wall, finished with paint or wallpaper.

WALK-IN CLOSET A closet large enough to walk into and store or remove clothing.

WALK-OUT BASEMENT A basement with a door on ground level allowing ingress and egress from the basement.

WALL COVERINGS Anything covering a wall, usually consisting of paint, wallpaper, drywall, wood, siding, and plaster.

WALL HUNG LAVATORY A bathroom lavatory designed to hang on the wall with no other support.

WASHER OUTLET BOX A metal or plastic box designed to be recessed in an interior partition, allowing the connection of washing machine water hoses and providing an indirect waste for the washing machine discharge hose.

WATER SAVER TOILET A toilet using three gallons of water or less each time the toilet is flushed.

WATER DISTRIBUTION PIPE The pipes carrying potable water to the plumbing fixtures.

WHIRLPOOL A trade term describing a bathing tub with whirlpool jets. These tubs are equipped with faucets and a sanitary plumbing drain. They are designed to be filled and drained each time they are used.

WHIRLPOOL JETS The devices found in whirlpool tubs and spas allowing jets of water to circulate the water contained in the bathing units.

WORKER'S COMPENSATION INSURANCE A type of insurance protecting workers who are injured while performing their professional duties.

WORKING PLANS Any set of plans adequate to allow tradespeople to perform their duties in a satisfactory manner.

Index

About the Author

Chase Powers is the author of several books on professional contracting. He is the author of McGraw-Hill's *Builder's Guide to Cosmetic Remodeling* and *Kitchens: A Professional's Illustrated Design and Remodeling Guide*. A licensed general contractor and a master plumber, he has 20 years' experience in both residential and light-commercial building and remodeling. His work has included all types of remodeling, both nonstructural and structural.